Planning, Measurement and Control for Building

Planning, Measurement and Control for Building

Robert Cooke

A John Wiley & Sons, Ltd., Publication

This edition first published 2009

Blackwell Publishing was acquired by John Wiley & Sons in February 2007. Blackwell's publishing programme has been merged with Wiley's global Scientific, Technical, and Medical business to form Wiley-Blackwell.

Registered office
John Wiley & Sons Ltd, The Atrium, Southern Gate, Chichester, West Sussex, PO19 8SQ, United Kingdom

Editorial offices
9600 Garsington Road, Oxford, OX4 2DQ, United Kingdom
2121 State Avenue, Ames, Iowa 50014-8300, USA

For details of our global editorial offices, for customer services and for information about how to apply for permission to reuse the copyright material in this book please see our website at www.wiley.com/wiley-blackwell.

Library of Congress Cataloging-in-Publication Data

Cooke, Robert.
Planning, measurement, and control for building / Robert Cooke.
p. cm.
ISBN 978-1-4051-9139-5 (pbk. : alk. paper) 1. Building–Superintendence. 2. Construction industry–Planning. I. Title.
TH438.C6458 2009
690.068–dc22 2009005454

A catalogue record for this book is available from the British Library.

Set in 10/12 Palatino by Aptara Inc., New Delhi, India
Printed and bound in Malaysia by KHL Printing Co Sdn Bhd

1 2009

Contents

Preface

As a companion volume to *Building in the 21stCentury* this addition should help explain the paperwork side of construction. It has been produced to help students who are on BTEC/Edexcel National Award/Certificate/Diploma, and Higher Nationals courses in construction who want to become architects, building surveyors, quantity surveyors, or construction managers etc. The book should also be helpful to advanced craft students studying for City and Guilds (C&G) or National Vocational Qualification (NVQ).

Opening with perhaps the most important document in construction: the contract. We look at the fundamentals of what many people pay little attention to until things go wrong. With examples of a basic contract through to the complexities of multi-million pound contracts, the chapter explains the reasoning behind preparing a good document. It includes, in contrast, consequences of real contracts that have ended up in court, highlighting the fact that contracts are not always foolproof.

Construction is definitely a team project, so we take a close look at all of the main team members and how they inter-relate as design teams and construction teams. Careers in the professions start somewhere, so typical educational routes have been included to help those who want to know more. Most people have heard of the architect and the quantity surveyor, but what do they actually do? What about all of the other consultants such as the structural engineer, the town planner, the building control officer and the army of technical people who actually design, plan and organise construction projects? We look at approximate costing, taking off, buying, setting up accounts, call offs and explain the terms and procedures that enable the clients to achieve their objectives. After looking at the teams we then follow the routes of the design stage and construction stage, observing how the teams inter-relate and how the various processes progress.

History may be dull to some but it helps to know why things have to be carried out in specific ways. Who owns the land? What is a deed? What is a covenant? Who makes the laws? Why have laws? What are the Building Regulations? Are Approved Documents laws? Did Building Regulations really start after the Great Fire of London? Hopefully you will find the chapter on Acts and Regulations useful.

The final three chapters follow three very different projects from inception through to handover, and all points between. Looking at real projects enables the pieces of the 'jigsaw of construction' to be put into place, starting with my hairdresser's shop (sorry 'salon') fit out. In contrast, we look at a new build small speculative housing estate with the issues of planning permission and meeting the Government's requirements on land usage. The concluding chapter provides a unique look at a very prestigious high tech office development in the heart of London. British Land Plc, the Mace Group of companies and Arup Associates gave permission to publish their Ropemaker office development as a case study. As you will see, the £400m-plus development is at the cutting edge of design and construction techniques, addressing the issues of sustainable construction using recyclable materials, and utilising natural rainfall, thus reducing the demand on precious drinking water. Solar energy panels and scientific principles are used to reflect natural light without solar gain, and duel fuel boilers have enabled the client to achieve a very low carbon footprint and a BREEAM rating of 'excellent'.

We are living in the 21st century and should be building for the future whilst meeting the demands of today.

Also available free is the Planning, Measurement and Control website (www.blackwellpublishing.com/cooke), where there are resources for students, teachers and lecturers, and at all levels of study.

Good luck with your studies! I hope this book will help you understand the paperwork world of construction.

We should never stop learning.

Robert Cooke

Acknowledgements

I would like to thank Dr Paul Sayer, Lucy Alexander and Richard Walshe plus the team at Wiley-Blackwell for their help and advice over the publication of this book.

I would also like to thank the following people and companies for allowing me access and permission to publish the chapter on their Ropemaker project; in particular Paul and Richard at British Land Plc., Paul and Lee at Arup Associates and Jonathan Foster, project director for the Mace Group of companies. As major commercial companies they were all keen to help students learn about the latest techniques and processes used in commercial developments. Thanks to the management team on-site for their time and help with the technical details, and a special thanks to Barry for the site tours throughout the contract, with and without the groups of students. Very special thanks to Beverley who orchestrated the myriad e-mails and set-up the various meetings – happy retirement.

Paul Dickenson	(Arup Associates)
Lee Hosking	(Arup Associates)
Richard Elliott	(British Land Plc.)
Paul Langham	(British Land Plc.)
Barry Beck	(Mace Group)
Darren Bedford	(Sense Cost Consultancy Ltd – Mace Group)
Dean Emblin	(Mace Group)
Garet Estensen	(Mace Group)
Jonathan Foster	(Mace Group)
Beverley Handley	(Mace Group)
Paul McDonald	(Mace Group)
Shaun Tate	(Mace Group)
Ashley Thorn	(Mace Group)

Thank you to the following people for their help and who ensured the information was up to date.

Davinder Khindley	(Asta Development)
	(Byrne Brothers)
Duncan Gaman	(Cityscape Digital Ltd)
Mike Stranks MRTPI	(Rochford District Council Planning Department)
Peter Myhill	(St Anns Builder's Merchants)
Annette Kirk	(SHE Software Ltd)
Brian McClave	(Site-eye.co.uk)
Roger Williams	(SGB Rovacabin Shared Services)
Stephen Astley	(Soane Museum)
Phil Bull	(Visqueen Building Products)
Philip Harris	(Wright Hassall LLP)

And very special thanks to Ann, my wife, who unlike a 'golf widow' has to put up with me at work all day and word processing at night.

Websites and further reading

www.adjudication.co.uk/cases/impresa.htm
www.arb.org.uk
www.architecture.com
www.architecture.com/Files/RIBAProfessionalServices/ClientServices/RIBA%20Outline%20Plan%20of%20Work%202007.pdf
www.astadev.com
www.astadev.com/software/powerproject/democentre/index.asp
www.bbacerts.co.uk
www.blackwellpublishing.com/cooke/teaching.asp
www.cibse.org
www.communities.gov.uk/documents/planningandbuilding/pdf/326679.pdf
www.communities.gov.uk/documents/planningandbuilding/pdf/planningpolicystatement3.pdf
www.communities.gov.uk/publications/planningandbuilding/creatingbetterplaces
www.communities.gov.uk/publications/planningandbuilding/ppsclimatechange
http://store.eleco.com
www.hm-treasury.gov.uk/d/80.pdf
www.hse.gov.uk/construction/cdm.htm
www.hse.gov.uk/forms/notification/f10.pdf
www.hse.gov.uk/pubns/indg383.pdf
www.ice.org.uk/homepage/index.asp
www.islington.gov.uk/DownloadableDocuments/Environment/Pdf/code_of_construction_booklet.pdf
www.istructe.org
www.jctcontracts.com/JCT/amendments.jsp
www.netregs.gov.uk/netregs/legislation/current/63616.aspx
www.opsi.gov.uk/ACTS/acts1990/ukpga_19900008_en_1
www.opsi.gov.uk/acts/acts1991/Ukpga_19910056_en_1.htm
www.opsi.gov.uk/acts/acts2003/ukpga_20030037_en_1
www.opsi.gov.uk/RevisedStatutes/Acts/ukpga/1984/cukpga_19840055_en_1
www.opsi.gov.uk/si/si2004/20042204.htm
www.parliament.uk
www.planningni.gov.uk/Devel_Control/Application_Forms/Forms.htm
www.planningportal.gov.uk
www.planningportal.gov.uk/england/professionals/en/1115314110382.html
www.ropemakerlondon.com

www.shesoftware.com/enterprise.html
www.soane.org
www.soane.org/soanebuildings.html
www.thameswater.co.uk
http://penelope.uchicago.edu/Thayer/E/Roman/Texts/Vitruvius/home.html
www.vam.ac.uk/collections/british_galls/galleries/54/index.html
www.wrighthassall.co.uk/resources/articles/art_construction07051.aspx

Further reading

Ashworth, A (2001) *Contractual Procedures in the Construction Industry*. Pearson Education Ltd, Harlow

Bickford-Smith, S and Wood, L (2004) In: Speaight, A. *Architect's Legal Handbook*, 8th edition. Architectural Press, Oxford

Brook, M (2008) *Estimating and Tendering for Construction Work*. Elsevier Butterworth-Heinemann, Oxford

Chappell, D (2007) *Understanding JCT Standard Building Contracts*. Taylor & Francis, Oxford

Chappell, D and Willis, A (2005) *The Architect in Practice*. Blackwell Publishing, Oxford

Cooke, R (2007) *Building in the 21st Century* . Blackwell Publishing, Oxford

Halliday, S (2003) *The Great Stink of London*. Sutton Publishing, Stroud

Lee, S, Trench W and Willis A (2005) *Willis's Elements of Quantity Surveying*. Blackwell Publishing, Oxford

Chapter 1

Building contracts

Perhaps the most important part of a construction project is the contract between two parties. It is often neglected until things go wrong. This chapter will hopefully throw some light onto the reasons for having a contract and why it is so important to read and understand them before signing.

The content and wording are both specific and should be written where possible in plain English to enable both parties to understand the meaning. Sometimes things will go wrong and the concluding section provides one set of consequences of a real case.

According to the Oxford Concise Dictionary the word 'contract' means 'a written or spoken agreement between two or more parties, intended to be enforceable by law'. The word 'contractor' means 'a person who undertakes a contract especially to provide materials, conduct building operations'.

Take a simple operation such as replacing a rotten timber window. Does the person who wants the rotten window replaced actually own the window and the structure? If they do, will they pay for the work to be done and when will they pay? What will happen if the person who wanted the window replaced changes their mind half way through the job? What would happen if the contractor decides he/she doesn't want to continue and finish the job? What will happen to the old window? Will it be left leaning against the wall or removed for disposal? As you can see with a simple job of replacing a rotten window there are a number of issues that could present a problem. If the contract has been purely word of mouth it may be difficult to prove who said what and what the agreement actually contained.

1.1 The building contract

On a small project a letter of confirmation of a verbal contract will give both parties a record of the agreement and an opportunity to see if something has been misunderstood. The letter, however, should contain relevant information such as:

- name and address of the contractor
- name and address of the client
- the date the letter was written
- the description of the work to be carried out
- the quality or the main materials to be used
- when the work is to be carried out
- how much the work will cost
- how much taxation will be applied and at what rate to the contract
- who owns the contractor – names of the directors
- what will happen to the waste.

Look at the letter shown in Figure 1.1. Is it a typical letter of confirmation? Does it fulfil the criteria? Now compare that with Figures 1.2 and 1.3. Although the letter is slightly longer it explains what the contractor thinks has been agreed and provides the client with an opportunity to see the agreement in writing.

There are still small issues that have not been covered in the specification. For example, what types of handles are on the opening windows? Will the handles have locks? Will they be white coated metal, polished brass or plain grey painted handles? Are there any stays to hold the opening windows at specific positions or will the openers be fixed to stainless steel friction stays as commonly used on upvc or metal replacement windows? Will there be a mastic bead sealing the edge of the window to the masonry to prevent rain penetrating the gap? What will happen if it is raining on the day the window

One Tree Hill Builders
The Oaks
High Road
Wenchester
Essex WN1 3PP

Dear Mr and Mrs Fields,

Thank you for your order.

To replace the window with a new double glazed wooden window 1500mm x 1090mm

For the sum of £670.30p

Yours sincerely

Figure 1.1 Typical builder's letter of confirmation.

is to be replaced? The contractor obviously is a tradesman and has specified that the window will be weathered by at least an undercoat of paint. The knotting has not been mentioned. Knotting is a sealing liquid, commonly shellac, that will prevent the sap in the knot from oozing through the paint finishes. The primer will prevent water or applied finish from soaking into the softwood timber. There are paints that the manufacturers claim are both primer and undercoat, but a two coat application will build up a better finish. The window can be fixed even in the rain and if required can be sealed with mastic when the weather is dry. The top/finishing coat can also be applied when the weather is dry. That then leads to the issue of payment. It is not the fault of the contractor that it was raining conversely the client has agreed to pay the contractor on completion of the work. In this case the contractor will have to return when the weather is dry and on a day that the client agrees to. In practice the contractor may come to another agreement and return after he has been paid. It all depends upon mutual trust between the client and the contractor.

Alternatively if at the end of the contract the work is only partly complete, the window arrived in a bare wood state and the fixer tried painting the wet

F.E.N.S.A Registration
No. ++++++++++

Tel: 0111 567 8910

One Tree Hill Builders
The Oaks
High Road
Wenchester
Essex WN1 3PP

23rd October 2009

Mr & Mrs J. Fields
39 The Lawns
Wenchester
Essex WNI 3PP

Dear Mr and Mrs Fields,

Thank you for your order for the replacement window as set out below.

Would you please review the information and if you agree sign and date the bottom of this letter and return one copy in the enclosed stamped addressed envelope.

1. To replace 1No. wooden window 1500mm x 1090mm to kitchen.
2. Replacement window in softwood to match existing window, primed plus one undercoat prior to delivery.
3. Bead and gasket glazed double glazed sealed units.
4. Frame to be mastic sealed all round internally and externally in white silicone mastic.
5. No responsibility can be held for internal plaster or wall coverings however the window installers will take care during the removal of the existing window.
6. Window to be finished with one finishing coat in brilliant white gloss.
7. Old window and all waste to be removed from site by window installers and work area to be swept clean.
8. Payment in full on completion of work.
9. Work to be carried out 5th November 2009.

For the sum of £670.30p plus VAT @ 15%

Yours sincerely

I accept this contract ....
Dated ....

Peter Hills (Dir) Derek Keen (Dir)

Figure 1.2 Improved builder's letter of confirmation.

window between showers then there will be an issue termed a dispute. The contractor will try to claim the payment and the client will not want to pay as the work is not acceptable as per the agreement. The client may need to have the work completed by another company who will also require payment. The dispute may end up in court. The contractor will need to show the court proof of the contract and the reason for going to court – the claim for payment

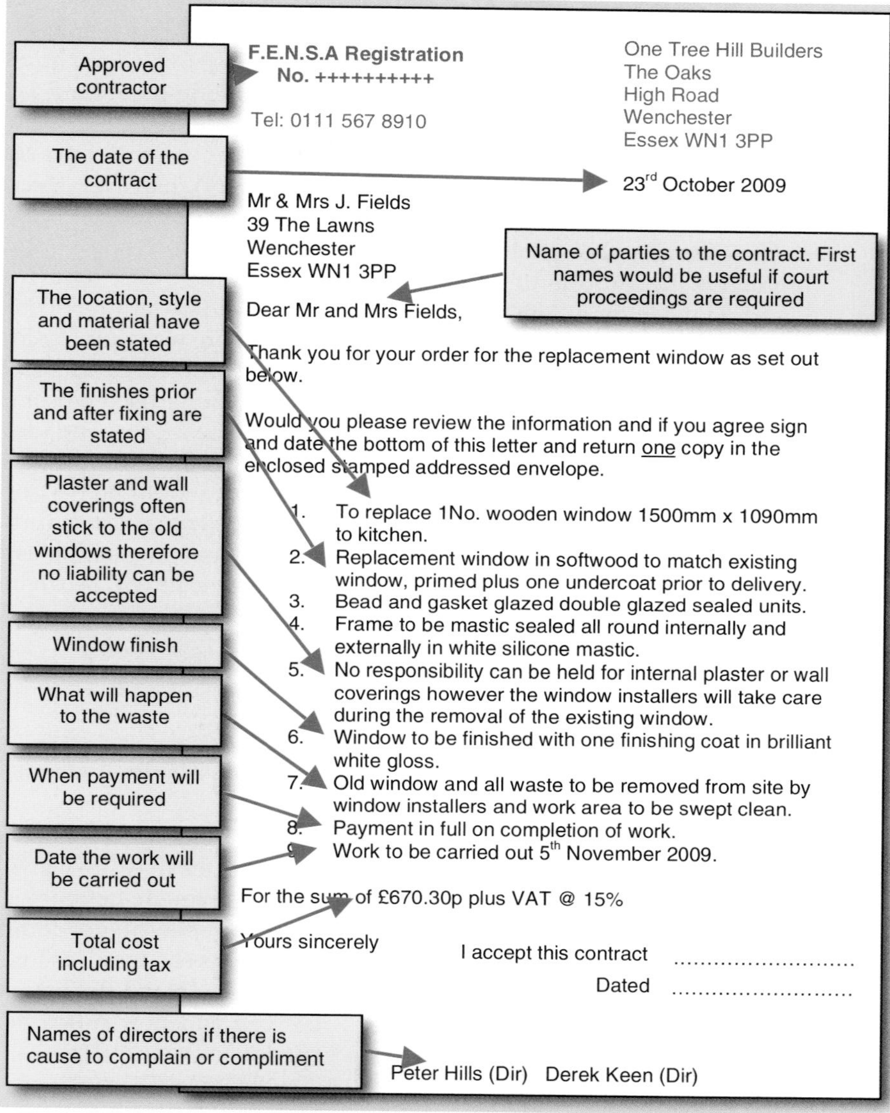

Figure 1.3 Figure 1.2 explained.

of work and materials. In this case a letter outlining the work etc. has been confirmed as accepted by the signature of the client. The client, however, has hopefully taken professional advice and taken photographic evidence including dates when the photographs were taken and the issues of the dispute. It would be useful for the client to employ a professional witness, someone who is qualified and experienced in the specific field of construction, for example, a qualified building surveyor or qualified building contractor. Notice the word 'qualified'. Anyone can call themselves a building surveyor, builder or building contractor. A window cleaner or site labourer can buy a van and a ladder and claim to be a builder. To become qualified in the subject, though, takes time and learning. You are probably studying to become qualified and know there is a lot more to construction than having a van and ladder, but there are plenty of unqualified builders touting for work.

The professional witness will provide a statement to the court confirming in their opinion the quality of the materials and workmanship. The court will then decide who is being reasonable and decide the outcome.

The example shown above is a relatively simple low cost contract but it highlights the importance of a written contract. If the project had gone ahead based on the letter in Figure 1.1 the problems would be far more complicated; worse still if it had all been word of mouth and there was nothing in writing.

Now let's look at the contractor's part of the contract. He has been approached by the client, perhaps as a result of an advertisement or the client has a friend whom the contractor has carried out work for. The idea of placing a sign on the side of the van and perhaps a sign board on site can attract business, but it costs money to advertise. The cost would be part of the contractor's overheads. It would not be practical to apply the cost against each job therefore overheads are lumped together and proportioned either as a time element, i.e. if the contract lasts one month the overhead will be divided by 12 and that portion added to the contract.

What other overheads are there?

1.2 Contractor's overheads

Premises to operate from are the first overhead to consider – small contractors may operate from their home. Perhaps their partner may answer the phone during working hours and therefore will require payment. Letters for business will require typing or word processing therefore a typewriter or computer will be needed together with stationery, stamps and pens etc. Invoices for materials and tools, known as plant, will need to be paid. The Inland Revenue will need to know how much the contractor has received and how much he/she has had to pay out, so records need to be kept for taxation purposes. Here are some further overheads:

- Premises to operate from – an office, a room in a house or – as with large companies – large buildings, workshops, maintenance bays, storage yards
- Services – heating, lighting, water, sanitation, ventilation, refuse disposal
- Standing charges – council tax, and utilities standing charges
- Office personnel – receptionist, accounts department, wages department, estimating department, buying department, planning department, transport department, and directors and their assistants
- Works personnel – foremen, chargehand, storeman, fitters (large companies may have their own garage workshop where company owned plant and vehicles can be serviced), workshop joiners, painters and so on.

- Plant – company-owned plant such as concrete mixers, adjustable steel props, hoists and so on, depending on the size of the company and the style of business. Some very large companies rent out plant to other companies. The advantage is that they can reduce their own costs for plant and maintenance, therefore helping keep their tender price low (see Chapter 6).
- Company vehicles – some companies lease the vehicles on long-term contracts whilst others prefer to buy the vehicles.

So far we have looked at a relatively small and simple contract. It would be impractical to, say, build a £100m office block using a simple letter of confirmation therefore more complex contracts are required. If you have ever had to make up rules for a local club you will know how difficult it can be. What you had intended to state can or may be interpreted by another person as something different. What about the issues that you have not included in the rules? Likewise laws (they are similar to rules) take learned people months or even years to decide on the actual wording. Is it practical for a contractor or an architect to spend enormous amounts of time designing a contract that covers all eventualities and can if need be tested in court? The solution since the end of the nineteenth century has been to use 'standard forms of contract' produced by the Royal Institute of British Architects (RIBA) known as 'the RIBA form of contract'. In 1931 two main groups, the RIBA and the NFBTE (National Federation Building Trade Employers), formed the Joint Contracts Tribunal (JCT). Advice from legal people, contractors, architects and other construction specialists enabled all facets of the new contacts to be represented. In 1939 and 1963 new contracts were developed to accommodate the new materials, methods and legislation. The Royal Institution of Chartered Surveyors (RICS) and other institutes have become involved to provide clauses specific to their members' needs. The outcome has strayed from the original ideal of the 'a standard form of contract' however the most recent changes have reduced the number of contracts available.

1.3 Contracts

In 2005 the Joint Contracts Tribunal (JCT) revised their standard forms of contract and further revised them during 2007 to produce nine new groups of contracts:

1. Minor works Building Contract
   - MW (Minor Works Building Contract)
   - MWD (Minor Works Building Contract with contractor's design)
2. Intermediate Building Contract
   - IC (Intermediate Building Contract)
   - ICD (Intermediate Building Contract with contractor's design)
3. Design and Build Contract
   - DB (Design and Build Contract)
4. Standard Building Contract
   - SBC (Standard form of Building Contract)

5 Generic contracts
6 Major Project Construction Contract
7 Measured Term Contract
8 Prime Cost Building Contract
9 Repair and Maintenance Contract.

These can be grouped into the following:

- MW and MWD
- IC and ICD
- SBC
- DB.

As with any new documents, they are constantly being revised to meet new regulations and provide improvement. The main revision relates to the changes in the Construction (Design and Management) Regulations 2007. For further detail look at the JCT website (www.jctcontracts.com/JCT/amendments.jsp) and www.hse.gov.uk/construction/cdm.htm.

Each of the groups of contracts has variants such as 'with quantities' or 'without quantities', and versions to include sub-contractors etc., the detail of which is outside the scope of this book. In addition to the four main groups there are specialist contracts:

- Short Form Sub Contract
- Sub-subcontract
- Major Project Construction Contract
- Measured Term Contract
- Prime Cost Building Contract
- Repair and Maintenance Contract for Commercial Work.

The new contracts can have additions or alterations made to them where a specific issue has not been fully covered due to the uniqueness of the project. However, if the contract changes are substantial they may fall into the domain of section 3 of the Unfair Contract Terms Act 1977.

As with the simple letter of confirmation shown in Figure 1.2, the fundamentals are the same:

- The names and addresses of both parties to the contract must be stated (where the parties have other business premises the address of the registered offices can be used).
- Each party should be identified: employer and contractor.
- The date of the contract must be included.
- What the project entails should be outlined (a description of the work including the location; the information would normally be in the form of contract documents).
- The cost of the work (usually itemised as per a priced bill of quantities or other priced documents – see Chapter 5).

Additional people will be included in larger projects therefore they should also be named:

- If an architect has been employed to design the project he/she will be named. There is usually an alternative given in the eventuality the architect dies.
- If a quantity surveyor has been employed (usually known as the PQS – private or professional quantity surveyor) to price the project on behalf of the client, they will be named. As with the architect, an alternative will be stated in the event of death of the QS.
- To comply with the CDM Regulations, a CDM co-ordinator must be employed by the client and will be named in the contract. It is common practice for the architect to either carry out the work of CDM co-ordinator or directly employ such a person.

The various sections shown above are termed 'articles' and are given numbers for ease of reference. The document becomes the 'Articles of Agreement' and would be signed by the parties entering the legally binding contract. As with the simple letter in Figure 1.2, the 'contract' is between two parties: the client, who is named the 'employer', and the 'contractor', the person(s) who will undertake a contract to provide materials and conduct building operations'.

In addition to the contract there are clauses that are designed to cover any eventuality not stated in the specification of the work. For example if the project becomes impossible to complete it is extremely costly and time-consuming going to court. Therefore a clause states that in that eventuality either party can have court action stayed (put on hold) and the dispute heard by arbitration. Arbitration is not legally binding, but both parties would normally agree to the clause unless otherwise stated. Arbitration would be carried out by well experienced professionals who have studied disputes as part of a qualification to become a Professional Arbitrator. They usually have a good knowledge of the legality of contracting and most of all the practical knowledge and experience of such works. Their function is to listen to both sides of the dispute and suggest an amicable outcome. Normally the action can be relatively quick and does not require the cost of barristers etc. to interpret the legality of such dispute. If after arbitration one party considers they have suffered damage they can still go to court and take legal action.

As previously mentioned, in addition to the contract a series of clauses will be attached. Their function is to clarify the multitude of events that may take place and what should be accepted or provided. There are several excellent books on the finer points of the clauses, but the following are some of the important ones:

- Completion date: When will the work be totally complete and will there be an opportunity for the employer to use part of the work? For example: the project is a supermarket and programmed to take eight months. The warehouse part is finished; can the employer start stocking out the warehouse part of the building whilst the contractor is finishing off the sales floor? This would be termed 'partial possession'. It should not be confused with 'phased' or 'sectional completion', which is more common on very large projects incorporating several buildings. Partial possession requires the employer to insure that part of the work and the contractor can reduce the insurance cover. However, it is open to problems. For example, the contractor cannot use that part of the site for storage or assembly work unless he has insurance cover.

- Changes to the design: Who can make changes to the design after the contract has been confirmed? These are covered in the clause 'Power to issue instructions' under clause 4: 'Architect's/contract administrator's instructions' as set out in JCT Standard Form of Building Contract. The architect can only issue instructions if express (i.e. specifically stated in writing) permission has been given. There are about 18 clauses where express permission has been stated, such as:
  - Clause 8.4.4 (inspections and tests): The architect can instruct the contractor to open up work such as a drainage run, or carry out an air test if he or she suspects the contractor may have used inferior materials or poor workmanship. If it is found that there is not a problem the contractor can apply for an extension of time, so inspections are not without penalty.
  - Clause 17.3 (rectification of defects): The architect can instruct the contractor to cut out damaged or defective materials. For example, if face brickwork has been damaged or defective facings to bricks have been used in a face finished wall.
  - Clause 36.2 (nominating a supplier): The architect can instruct the contractor which supplier will be used for a specific material, for example, where the colour of a sand cement render can only be achieved using sand from a specific supplier.
- Certification that the work has been carried out correctly: Certain contracts contain a clause that gives custody of part or parts of the work to the employer. For example, the architect will inspect a completed part of the contract and issue a certificate stating that the work is of the required standard. These are 'interim certificates'. When the architect signs off that part of the work as complete, the contractor will be able to claim payment within 14 days. Interim certificates are covered by Clause 30: Certificates and payments. The architect is required to issue certificates at times specified in the appendix of the contract. The PQS or cost consultant will normally authorise payment within 14 days of the certificate being issued. If the payment is late then the contractor is entitled to claim interest on the sum.
- Statutory obligations: Who should inform Building Control about the building work? Who should let the Health and Safety Executive know about the work? Who should apply for licences to erect a hoarding on a public footpath, suspend parking bays, divert traffic and so on? The contractor has that responsibility under the clause 6: Statutory obligations, notices, fees and charges.
- When a contract commences, the employer hands over the site to the contractor, termed 'taking possession'. The employer still owns the property, but the site becomes the legal obligation of the contractor. The site must be insured. The contractor must carry public liability insurance as a minimum, plus professional indemnity insurance. If an incident happens on the site it will be the contractor that is liable. If the incident is, say, a serious or fatal accident then the Heath and Safety Executive will be

involved and where applicable it is the contractor that can face prosecution under the Health and Safety at Work Act 1974 and/or several other Acts relating to safety issues on site. Insurance is covered in three clauses; Insurance against injury to person or property, Insurance of the works, and Liability of employer.

There are many more clauses; some pertain to all of the contracts whilst others are either omitted or reduced in detail. To understand contracts is a specialist skill and should be left to either legally qualified or professionally qualified people. Signing such documents is easy. However, ignorance of the law is no defence. There are several excellent books that give good guidance, but they cannot provide experience and qualification.

1.4 Client Management Contracts

Over the past years, since about the early 2000s, large development companies have used 'Client Management Contracts' in favour of 'Standard Building Contracts'. The costs and risks of building high value structures has become too much for many contractors. Projects that run in to multi-millions of pounds and over several years are by their very nature risky, especially in the current financial climate where some of the largest banks, financial and insurance companies in the world have collapsed (September 2008). Major contractors have become reluctant to even tender on SBCs.

In contrast a Client Management Agreement (CMA) places the risks back on to the client. What are the risks? Using a SBC the main contractor is responsible for the full procurement of the project including delays, and fluctuating costs. CMAs are not without risk though. In one case, the client, a consortium of hoteliers, entered into a CMA as Great Eastern Hotel Co. Ltd, with John Laing Co. Ltd. as the principal contractor (PC). The project was a major refurbishment of the Great Eastern Hotel in Liverpool Street, London. The project sum was about £35m including a contingency (an amount set aside that can be used if needed). However, the final cost ended up in excess of £61m. The subsequent legal action Great Eastern Hotel Co. Ltd v. John Laing Co. Ltd. [2005] was based on a 'breach of obligation' on behalf of the principal contractor. The client successfully sued on the basis that the PC had not ensured that the programme of work by the trade contractors was workable with a consequence of late completion and running over budget.

According to the Contract Journal.com the CMA had clear clauses set out in the schedule to the agreement. For example: The construction manager:

- 'shall procure that each trade contractor complies with all its obligations under their respective trade contracts' and
- would 'provide such management, control, administration and planning with the work of the trade contractors so as to ensure full compliance by the trade contractors with their respective trade contracts'

Laing in defence claimed they were contracted for the provision of services and therefore had no responsibility for the trade contractors. On this point Judge Wilcox agreed. However, Laing had failed to manage the various trade contractors and had not ensured they completed their contracts on time. The delays had consequences as knock-on effects, to mitigate which the PC had tried to reprogramme the project and claw back lost time.

Chapter 2

The design team

Logically the project will need designing and planning before putting into the operation of construction. This chapter will introduce the design team and working relationship of the network of professionals involved in the design process.

For students intending to enter the professional side of construction explanations have been provided illustrating the input that each team member will offer. To further help with careers information, pathways of study and academic requirements to become qualified have been given. Construction is very much teamworking although who receives the accolades does not always reflect the fact.

Building is often referred to as a game; the building game. If compared to say a football match there are two teams, the design team and the construction team. The overall outcome hopefully will be a satisfied client, but how does that happen? Looking at the design team first, they will comprise:

- client
- lead consultant – (architect*)
- consultants
  - private quantity surveyor
  - structural engineer
  - building services engineer
  - fire engineers
  - town planner
  - Local Authority planners
  - architectural technologist
  - services engineer
  - civil engineer
  - building surveyor
- specialists
  - manufacturers and suppliers.

* The 'architect' could be replaced by others such as architectural consultant, designer, architectural technologist etc. However, there are legal restrictions for a person using the title 'architect' – see Architect.

2.1 Client

The client can be a solitary person who wants a specific structure to be built or other construction project such as a new dwelling, refurbishment or extension to an existing building. Alternatively the client could be a group of directors who require new business premises. In either case the client is the person or persons who instigate the work.

2.2 Architect

Commonly the lead consultant, the title 'architect' is protected by the Architects Act 1997. To identify yourself as an architect you must be qualified to a required standard and be registered with the Architects Registration Board (ARB). It is illegal to call yourself an 'architect' unless both qualification and membership are complete. To become qualified there are several routes starting with either GCSEs or A/AS levels. The ARB has an excellent website providing the latest guide to the universities that offer relevant architectural qualifications; see www.arb.org.uk. Alternatively, look at the Royal Institute of British Architects (RIBA) website: www.architecture.com/.

Generally a student should start with suitable 'A' or 'AS' levels or 4 GCSEs at C grade or above in:

- English
- Mathematics
- Science
- plus either Geography or History.

Obviously, art-related GCSEs or 'A' levels will be useful. However, they should be in addition to those shown above. Architecture is not how 'pretty' or how accurately someone can draw. Architects need to be able to understand science, be able to read and write, and have an appreciation of what has already been built, therefore history and geography will be useful. Historically, architects would travel gathering inspiration from the works of others. Those who could not travel far would trade ideas from sketches and discussion. It is thought that Sir Christopher Wren based the dome on St Paul's Cathedral on the dome of St Peter's in Rome (see Figure 2.1). Wren was a professor of astronomy, an inventor, a mathematician and an engineer. When he was young it is thought that he acquired his interest in Roman architecture after reading the book *De Architectura* by Marcus Vitruvius Pollio, a Roman architect who lived in the 1st century AD. For an insight into the book it is worth looking at an excellent website by Bill Thayer that provides an English translation of the works. Several translations have been produced over the centuries, but Bill has chosen that of the London architect Joseph Gwilt. Gwilt was famous for

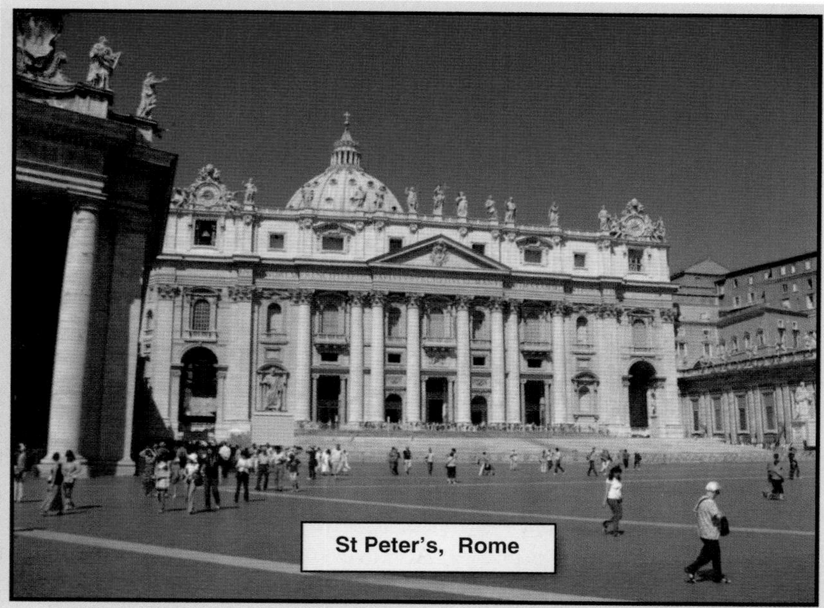

Figure 2.1 The basis for St Paul's cathedral in London.

his involvement in the Metropolitan Building Act 1855 and the development of Victorian London.

- http://penelope.uchicago.edu/Thayer/E/Roman/Texts/Vitruvius/home.html

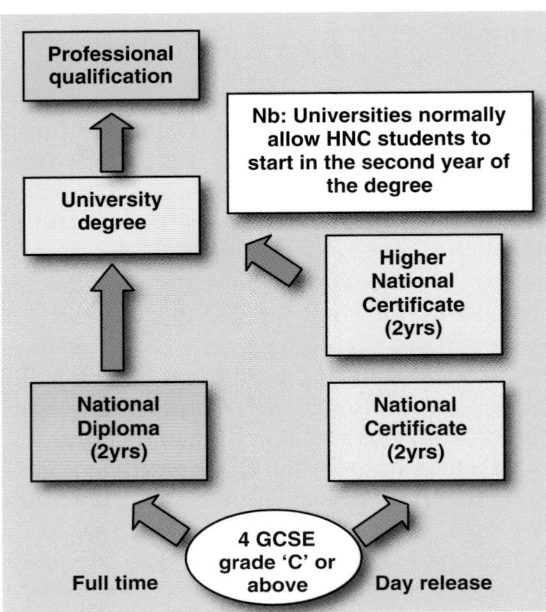

Figure 2.2 The Further Education route to becoming a qualified building surveyor.

Returning to the subject of education, many FE (Further Education) colleges offer BTEC courses in Construction such as the National Certificate (NC) and Diploma (ND), leading on to the Higher National Certificate (HNC) and/or Diploma (HND) (see Figure 2.2). At the time of writing it is anticipated that a new range of BTEC Diplomas will eventually replace the existing educational courses. After achieving either the National Diploma or HNC many students progress to university to study at degree level. Students can study full-time, enabling more units per year to be studied, or attend one day a week over a longer period (see Figure 2.3). The advantage of having a job and day release is that it provides experience in the real world where problems have to be overcome, all the while studying the theories of construction. Colleges provide a range of experiences. For example, a student may work for a small architectural practice specialising in 'one off' houses. He or she will gain knowledge whilst at work but is unlikely to ever see or learn about foundations, walls, sanitation or any construction science. Another student may work for a quantity surveyor and never see a building site, so the one day at college will provide a wide range of experiences. In contrast, a full-time student may achieve the entry qualification more quickly with a more intensive programme of study, but work experience and field trips are an important source of real world experience.

Academia (theoretical study only) is only one part of the study programme. Historically the student had to pay the 'master' for the experience and education. Eventually apprenticeships developed enabling a contract between an employer and a student where the student no longer had to pay the master but received a weekly wage. The system worked very well whilst companies and firms directly employed people, but it is not as common now, as many employers sub-employ (see Section 3.1, Trades).

School leavers who wanted to join the professional side tended to have GCEs or CSEs (the forerunners to the GCSEs of today). There are two distinct paths in construction: the 'professions' and the 'trades'. When I left school,

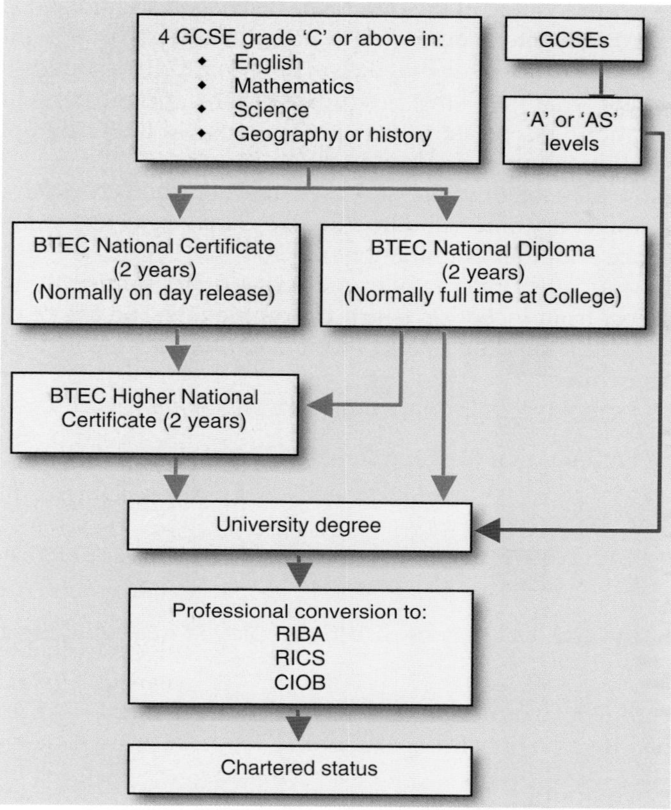

Figure 2.3 A more detailed academic route to becoming professionally qualified.

those who were good at maths and English tended to go on to University or join a bank or insurance company. Those who were less academic learned cookery (or domestic science as it was commonly known), gardening, woodwork or metalwork and went on to work in shops or 'get a trade'. There are still many careers advisors and schoolteachers who believe that school students who are more able academically should stay on at school to gain higher grades. Those who are less able or perhaps disruptive in school should go to FE colleges to further their studies in 'vocational' pathways. It should not be the case though. As long as there is an interest in the subject, students should follow the best route of education for them, not to be just a statistic on the school educational league tables.

Interviewing prospective students wanting to join an FE college on the 'professional-side courses', an all too common problem is: 'I'm good at drawing and my careers teacher said I should be an architect'. Unfortunately there are still 'advisors' out there that have little or no knowledge of what the professional side of construction is all about. Many schools want to keep their brighter students as it is financially beneficial and looks good for the school's achievement record. Hopefully in this section of the book you will be able to make your own mind up, based on facts.

It might be worth looking at the difference between education and training. You can train a dog to fetch your newspaper but the dog may not understand why. Likewise a CAD operator may know how to produce superb drawings but has not a clue what they have produced, or why specific parts have been selected. Education in contrast to training will enable the CAD operator to know what components go where and why.

Most of the professions require seven years of study to become qualified (five years at university on a validated degree course plus two years' professional experience and final examinations).

The job of the architect can range from 'design only' through to overseeing the whole project from inception (start) to completion. The extent of the services being offered should be set out in the contract documents between the client and the architect.

The architect can work in 'private practice'. That is

- working on their own (self employed)

or

- working with partners or associates as part of a firm or practice of architects.

Alternatively, the architect may be directly employed in a commercial organisation:

- many commercial outlets such as a bank or chain of fast food restaurants have their own architects department

or as part of an authority:

- Local Authorities such as County Councils, Borough Councils, Town Councils and District Councils may have their own architects department. However, with ever-diminishing funds they now more commonly put their work out to private practice.

Commercial architects, although still needing to be fully qualified and registered with the ARB to use the title, are not in the same position as those in private practice. They are in effect employees of the company whereas the architect in private practice has the option of whether to take on work or not.

Large companies such as supermarkets chains, fast food restaurants and multiple retail outlets may have their own architects departments. The advantages are that a corporate image can be designed into each building. The layouts of the stores can be designed to enable the customer a feeling of familiarity irrespective of where the outlet is. A fully qualified architect may head a team of architectural technologists who provide the parameters of design criteria for 'contracted architects'. In effect they act as the client.

Private practice architects are 'contracted' and usually work on behalf of the client (see Figure 2.4a). They are in effect employed by the client in a professional capacity. As with all contracts, they can be terminated before completion therefore it is important for both the client and the architect to mutually agree to the contract (see Section 4.1, Choosing an architect).

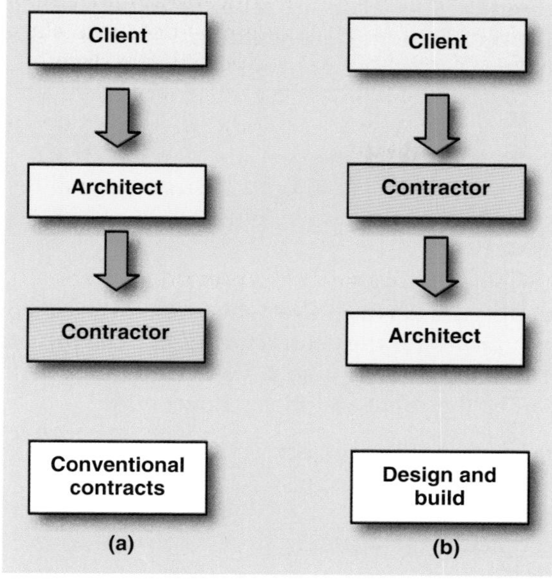

Figure 2.4 Contractual alternatives.

Alternatively, the contractor may employ the architect (see Figure 2.4b).

Only one architect can enter into the 'contract' with the client. If the project is of a sufficient scale other architects can work under the contracted architect or the project can be divided into several contracts. For example, the client may commission an architect to design a shopping mall. The contract may cover the structure finished as a shell. Other architects may enter into contract with either the main client or their letting agent to fit out the shop units and offices where applicable. Each architect would be contracted with a client or agent of the client but would not overlap in their responsibilities.

The prime objectives for the Architect are:

- to fulfil what the client 'wants' and identify the client's 'needs'
- to provide aesthetic value to the project – what the finished project looks like
- working operation of the building – the functional requirements.

It is important for the architect to convert the requirements of the client into tangible formats that others can work to. If the client wants a tall blue glass-walled office building, circular in plan, it is not up to the architect to design a brick-clad rectangular low-rise structure. If the architect does not want to design what the client wants then they should not enter into the contract. However, what the client 'wants' and what the client 'needs' may not be the same, and the needs of the client should be considered.

The professional experience of the architect will be based on the study of both aesthetics and functional use of the proposed structures. The experience of the architect is based on feedback from previous projects and should be of particular benefit to the client. It should be part of the selection process (see Section 4.1, Choosing an architect). Buildings are individual creations. Unlike, say, a car, where the client can see the finished work before committing to a contract to purchase, a building contract will be signed before the many hours of discussion, design and building and can only be fully assessed when complete.

An experienced architect will have made mistakes or worked on projects where the design may have looked good but functionally did not work. That is the process of learning. However, the client will not want to finance a project that may or may not work, therefore the selection of the architect should be thorough.

As buildings become more complex, specialists are required to help design specific aspects of the project. Commonly termed 'consultants', they form part of the design team.

2.3 Consultants

Private quantity surveyor (PQS)

There are two main teams of quantity surveyors; those who work for the client (PQS) and architect and those who work for the contractor (CQS). The private quantity surveyor handles the costing for the project. Working closely with the architect they provide and assist the client and the architect with:

- approximate costings at the outset of the project
- more accurate costings for the tendering stage
- taking off the quantities of materials
- preparing the contract documents
- assessing the tenders
- agreeing payment to the contractor.

The architect will convert the client's requirements into a series of hand-drawn sketches to enable the client to see the scheme. With today's technology there are CAD (Computer Aided Design) architectural packages that enable fluidity of design. The operator can extend, reduce, invert, change the rendering etc. Other packages enable virtual reality that allows the client to enter the electronic project. The software enables computer generated images of the finished project so that the client can virtually walk round the building, viewing 360° in both planes. Some software even allows reflections and shadows, giving even more realism. However, hand sketches still have life and beauty and creativity, something a machine still lacks – or is that just me?

When the client agrees with the overall sketched design, the PQS will be able to provide a guide figure (cost) for the project. In a similar way to choosing a dentist or hairdresser, the PQS would be consulted for their experience. The PQS should have a portfolio of work that is of a similar nature to the proposed project. It would be very unwise to select a PQS who had only previously worked on low-rise housing if the project is a shopping mall.

At this stage the client and the architect would have no idea of the cost of the proposed project. The architect would have asked the client what budget had been given to the project, but neither the architect or the client would know how much the project is likely to cost. Therefore an 'approximate costing' would be calculated by the PQS based on previously completed work of a similar nature. It is a 'guesstimate' intended as a rough approximation of the final cost, in a similar way to someone saying 'it is about 200 metres on the right'. They do not really mean they have measured the distance or that it is 200 metres as opposed to 199 or 201 metres, the intention is a guide only and they have 'guessed' the approximate distance. Commonly the PQS would base his/her approximate costing on one of the following methods:

- a cubic method – price per cubic metre £/m^3
- an area method – price per square metre £/m^2
- a unit method – price of each item (manhole cover, door handle etc.)
- or a lineal method – price per linear metre £/m.

For example the proposed project is a shopping mall. The architect has established the approximate overall dimensions, how much commercial floor area (office space), retail floor area (shops) and in this example residential space (apartments – similar to flats, but using the word 'apartment' will make them more desirable and therefore increase their value). The PQS will look at any similar projects that they have been consultants to that have been completed in the last, say, five years. The completed projects would provide actual final costings for everything other than the price of the land. Land costs can fluctuate depending upon where it is and the use the land can be put to. For example, 4 hectares in an area that has been neglected and has high unemployment will be less attractive for a shopping mall than, say, in or close by a vibrant town where wages and employment are very good.

The PQS will if possible look for alternative forms of construction to contrast costs. For a shopping mall a steel-framed superstructure with metal-clad walling or a cast in situ concrete-framed superstructure with masonry walling could be used. At this stage the client hopefully will know the external aesthetics but cost will obviously be a factor. The PQS would be able to break down the previous costing to a price per metre square for the cladding and the superstructure framework may be taken back to a volume cost. The architect at this stage would have established the quality of the proposed project, so if the completed projects included high value features or finishes then they would be removed from the total.

Example

A completed shopping mall has a glass domed roof over a large atrium (an atrium is similar to a hall or well that allows natural light within the building almost like a covered courtyard – see Figures 2.5 and 2.6). The client on the proposed project wants the maximum floor areas therefore the design will not have the atrium. The PQS will extract the cost of the glass domed roof and calculate the cost of a square metre of retail shell.

Say the structure area has a footprint of:

$$300\text{ m} \times 500\text{ m} = 150,000\text{ m}^2$$

and has four storeys (floors including the ground floor). Therefore 150,000 × 4 = 600,000 m^2 floor area in total.

Say the cost of the superstructure was £60,050,000. The glass dome cost £50,000.

Superstructure	£60,050,000
Glass domed roof	−50,000
	60,000,000

Cladding to external surfaces with a storey height of 5 m therefore:

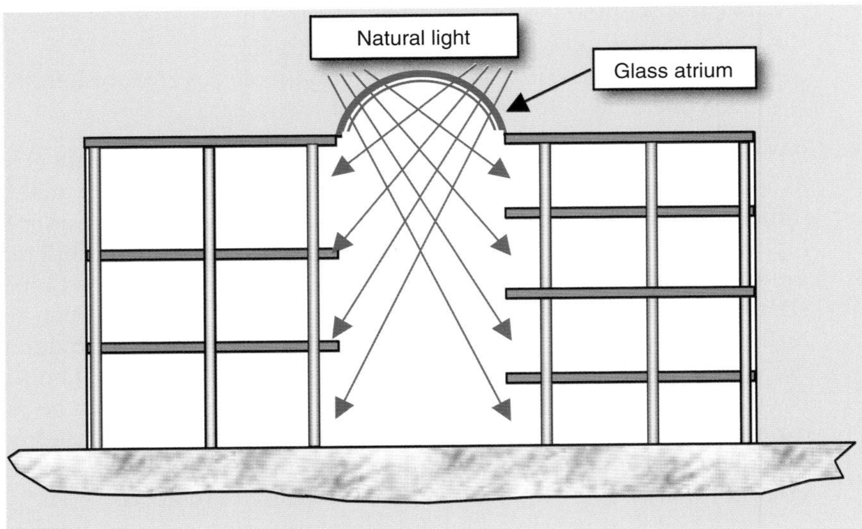

Figure 2.5 Section through a glazed atrium.

Figure 2.6 End elevation of a glazed atrium.

Length × height (storey height × number of stories) × number of walls

$$(300 \times (5 \times 4)) \times 2 = 12,000 \text{ m}^2$$
$$(500 \times (5 \times 4)) \times 2 = \underline{20,000 \text{ m}^2}$$
$$32,000 \text{ m}^2$$

The cost of the cladding including fixing will probably have been priced or 'billed' (a term used for anything written in a bill of quantities) as a 'lump sum' by a specialist sub-contractor and therefore be relatively easy to subtract from the total sum. Subtract the cladding cost from the total sum. More commonly the lump sum will be in the form of a tendered sum from a nominated sub-contractor. The cladding system would be chosen by the designer and an approximate price obtained to ensure it will be within the budget. The designer may at this stage consult the client for comment and/or approval, or wait until the whole design package is complete. When the client has given approval the designer or PQS will ask the cladding sub-contractor to provide a quotation for the work that will be written in the bill of quantities as a lump sum. Alternatively if there is more than one cladding contractor who can provide the same work the may be asked to tender for the work. Where the contract has a bill of quantities the nominated sub-contractor will have their company name, address and telephone number written in next to the work;

> 'Saxon Aluminium' Glass and glazing external cladding as supplied and fitted by: James Clark and Eaton Ltd. Unit 32 Perry Road Industrial Estate, Witham, Essex CM4 3UL. O123 456789.

When the bill of quantities is eventually sent out for pricing the estimators will be able to contact the nominated sub-contractor who will provide the lump sum to enter in the bill. If the estimator then goes to another company for a cheaper quote it would not be valid as they would not have used the nominated sub-contractor. The main reason for nominating the sub-contractor is to ensure the quality and back-up both in technical and financial terms.

An example was a prestigious office development in Ipswich. The company I work for had spent months working with the architect to provide a design that would allow the continuous glass façade to expand and contract as the solar heat went around the building. The architect then used the knowledge and the design drawings as tender documents. A relatively small company without any design offices or technical representatives, thus much lower overheads, put in a far lower quotation and won the work. However, they then contacted the company I was working for to find out where they could procure the special glass fittings and how to actually glaze the project. (You may guess what answer they received.) From that day on no technical representatives would enter into design work or send work in to the design office unless a 'letter of intent' had been issued by the architect or designer. A 'letter of intent' is a form of contract stating that the designer (who is the client) agrees to pay the contractor for work carried out as requested. Then if the designer wanted to use another cheaper contractor to actually carry out the work the designer would have to pay for the technical expertise of the first contractor.

If the project is based on a Client Management Agreement without a bill of quantities then the sub-contractor for the cladding would enter into a CM/TC Client Management Trade Contract directly with the client (see Chapter 10).

The lump sum for the cladding had been £6,400,000.

Therefore £6,400,000/3,2000 = **£200 per m^2 cladding as fixed**

Shell including cladding	£60,000,000
Cladding	– 6,400,000
	£53,600,000

Continued

Superstructure total floor 600,000 m^2
Therefore £53,600,000/600,000 = **£88.33 per m^2 floor area**.
The proposed shopping mall will have five storeys with a total floor area of:

$$250 \times 300 \text{ m} \times 5 = 375{,}000 \text{ m}^2$$

Therefore 375,000 m^2 @ **£88.33 per m^2** = £33,123,750
Plus external cladding:

$$(250 \times (4 \times 5)) \times 2 = 10{,}000 \text{ m}^2$$
$$(300 \times (4 \times 5)) \times 2 = \underline{12{,}000 \text{ m}^2}$$
$$22{,}000 \text{ m}^2$$
$$22{,}000 \text{ m}^2 @ £\mathbf{200} \text{ per } \mathbf{m^2} = £4,400,000$$

Approximate cost for the proposed shell will be:

$$£33{,}123{,}750$$
$$\underline{+4{,}400{,}000}$$
$$37{,}523{,}750$$

In addition to the cost of the shell office accommodation and apartments will need to be added. In a similar way based on 'fitting out' similar work in framed structures the PQS will be able to extract a cost per m^2 floor area and make adjustment for the external cladding which obviously will require glazed areas.

The PQS has projected approximate costs that include all materials (delivered or removed) and the labour and plant required to carry out the project. Costs for the site preliminaries, foundations or basements will need to be added to the approximate costs.

When the project has been designed and permission obtained to build the project the architect will produce detailed drawings known as 'working drawings'. The drawings will be precise showing dimensions, the materials and finishes, and relevance to the project. For example plans (that means looking down on the project directly from above). There will be several plan drawings:

- site layout
- foundation layout
- drainage layout
- ground floor layout – position of walls, windows, doors, lift shafts, stairs etc.
- upper floor layouts – position of walls, windows, doors, lift shafts, stairs etc. (typically one layout per floor)
- roof layout
- services layouts.

In addition there will be sectional drawings. They are drawings showing mainly vertical sections through such things as storey heights (floor to ceilings are termed storey heights) wall structure, windows, lintels, beams or joists.

- foundations section
- ground floor section

- upper floor section
- roof section.

Elevational drawings show the outside faces (elevations) of the building. Commonly compass directions can be used or reference to important features such as:

- North elevation
- South elevation
- East elevation
- West elevation

or

- High Street elevation
- front elevation
- rear elevation
- river elevation.

Where important details need to be enlarged such as special fixings to window sections, masonry fixings, frame components etc. then they are normally identified on the main drawings and referenced as 'Detail 01 – see drg no. 00124'. They may appear in plan or section commonly to a scale of 1:50, 1:20, 1:10 or 1:5.

The PQS will use all of the drawings to carry out a 'take off' to quantify the amount of materials required and prepare a list for costing termed a 'bill of quantities'. Basically it is a shopping list itemising all of the materials required to complete the building in a similar way to writing a list of ingredients to make a cake. Note the list will not show the machinery or number of people required, or how much wastage or how to build the project. To help simplify the lists similar materials and or trades will be collected as suggested in the Standard Method of Measurement (see Chapter 5).

The PQS will prepare estimated costs for the materials and add lump sums where individual items will be supplied as a complete package. For example brickwork could be itemised as bricks, mortar, ties and reinforcement as separate quantities, or cost per m^2 as built. In contrast 30 lighting posts from a specific manufacturer would be billed as one price – lump sum: 30No. lighting posts [LP1234-grey] delivered to site, off-loading by others – for the sum of £3000 plus VAT @ 15%.

Where items are supplied and fitted by a specialist company the PQS will have a quotation from the specialist and the item will be billed accordingly.

The PQS will need to know how to read and interpret the drawings from the architect and other consultants. (The drawings must be via the architect to ensure they are correct and as required.) This is the main reason for student quantity surveyors having to attend college and university to study construction. Quantity surveying is more than handling money. To become a quantity surveyor a similar route to that of the architect is followed up to BTEC/Edexcel National Certificate in construction. Some FE (Further Education) colleges offer an HNC (Higher National Certificate) based on more measurement units targeted on quantity surveying and estimating. Alternatively there are specific quantity surveying degree courses offered by other colleges and universities. If you intend to follow through to become a member of the RICS (Royal Institution of Chartered Surveyors) it is very important to find out which units and degrees will meet their requirements. This is true of all the professions. Ensure the course subjects are the right ones for progression otherwise you may be wasting your time. An HNC in construction, though, will provide an excellent grounding for all the construction professions. A QS will benefit from knowing the basics of demolition, services, design procedures and so on. If, for example, a student leaves school with 'A' levels and goes directly into university on a quantity surveying course they will miss out on many of the other parts of construction education.

Structural engineer

Commonly a consultant to the architect, the structural engineer would look at the proposed design from a 'practicability of construction' point of view. The architect will provide an artistic approach to the clients' idea but the structural engineers will look at the design of the structure.

Example: The design of a wall

From the architect's perspective – functions of the wall:

- Defines the area of a room.
- Is able to take hanging wall cabinets (important if the wall is studwork).
- Colour and texture.
- Finishes.
- Will it satisfy the fire requirements of the building/element?

From the structural engineer's perspective – the wall as an element:

- Will the wall be loadbearing?
- Is the wall stable – will it require buttressing or retaining?
- What minimum compressive strength will the wall components need to be? If it is to be aircrete such as Thermalite blockwork, will, say, a Turbo block be strong enough? Or will a high strength block be required?
- What type of mortar and mortar strengths will be required?
- Will there be other loadings on the wall such as live loads from a floor?
- Will the wall be subject to wind loads?

As you can see there are many questions of design for such a simple element. Many architects will actually take the above into consideration and the structural engineer will check the information. When a structure becomes more complex such as the famous Lloyds building, and Canada Tower, the structural engineers will have a stronger input in the design.

To become a structural engineer a good ability with numbers and problem solving is required. Structural engineers need to have a good understanding of mathematics and physics as well as construction design. The academic route would be the same as that for the architect but diverging at degree level. To find out more about structural engineering visit the Institution of Structural Engineers: www.istructe.org or the Institute of Civil

Engineers: www.ice.org.uk/homepage/index.asp. Both institutes encourage student membership and guidance for a career in structural engineering.

So far the design team comprises:

- architect
- PQS
- structural engineer.

Between them they have converted the clients' idea into a series of graphics and provided an approximate cost for the project. The type of soil will dictate the foundation design therefore an approximate cost for the substructure can be projected.

The approximate cost should be within about 25% of the actual cost or the project.

Building services engineer

The building will need lighting, heating, cooling, water, sanitation, electricity and perhaps gas. They are the services that on a domestic property would be designed by the architect. However, on large, more complex projects like hospitals, office developments, shopping malls the services need specialist detailing. On larger projects the electrical cabling and switching become particularly complex, also ensuring water pressure is adequate on the top floor of a multi-storey building. Lifts and escalators may be required therefore it is wise to involve a building services engineer at a very early stage of the design. As a consultant they will provide initial design input requirements that will help the architect and structural engineer even as early as the feasibility stage.

Architects and the structural engineers have designed buildings and then had to make significant alterations to facilitate the size ducting required to air condition the building. All time and money wasted and a problem that would have not existed if the building services engineer had been involved earlier.

The building services engineer will calculate pipe sizes and run lengths for fluids, air ducting sizes and shapes, boiler sizes for heating water, space heating plant, lifts, escalators, fire fighting and fire prevention equipment.

The architect may choose a company which specialises in, say, passenger lifts. The company will provide the design of the cars (the container you stand in whilst being lifted), lifting mechanism and plant requirements, but the electric supply cable and switching gear required will be designed by the building services engineer. The structural engineer will look at the loading implications, the positioning of the plant room and the stability of the lift shafts. As you can see, design is certainly 'teamwork'.

With the introduction of more stringent controls on carbon output from buildings building services engineers would be involved with the solar gain, shading and heat loss of buildings. Until recently the housing market had relatively simple calculations based on heat loss through the elements (walls, windows and doors, roof and the floor. The heat loss could be calculated on the 'elemental approach' where trade-offs between elements could be achieved as long as the overall loss and heating supply remained equal to or less than that stated in the Building Regulations under the SAPs scheme (Standard Assessment Procedure for energy rating of dwellings). The amendments to the Approved Documents L require specialist involvement therefore building services engineers are more involved in the design.

Sound can be a nuisance when it is unwanted, therefore sound insulation of buildings can be complex. Filtration and direction of sound in buildings such as lecture theatres and large halls requires specialist skills and know-how. DPD Associates (architects) partnered with Salford University Acoustics Research Centre spent £60 million on a refit to improve the acoustics of The Royal Albert Hall.

The career path for building services engineers would be similar to the architect, however, some colleges offer different BTEC units specifically concentrating on the services. Progression would follow with a specialist degree and professional recognition with the Chartered Institution of Building Services Engineers (CIBSE). Their website is www.cibse.org.

Fire engineers

Fire engineering can be very complex. New buildings and alterations to buildings require compliance with the minimum standards set in the Building Regulations and Approved Document B: Fire Safety. The underlying requirement is for the safe evacuation of building occupants, the control of the spread of fire internally and externally, and to meet the needs of the fire service for access and fire fighting. The Approved Document B comprises two volumes: domestic and non-domestic. They have been updated in April 2007, but it is good policy to check the Government website to check for the latest upgrades and sheets of erratum (errors) (www.planningportal.gov.uk).

As previously mentioned, the regulations are the minimum requirement. Fire engineering should be based on practical experience and commonly ex-fire fighters/officers become fire engineers when retired from active service.

Fire is commonly associated with flames. However, smoke or fumes are responsible for most deaths. One material may have a burning characteristic tested in laboratory conditions but when placed in specific conditions with other materials can behave in very different ways. Examples of tragic fires that have influenced change in the fire regulations include the Stardust Nightclub in Dublin (1981) where 48 people died and over 200 were injured. The material used had been tested and complied with the regulations but they had not been tested together, where their fire behaviour was notably different. A few years earlier in Summerland Leisure Resort on the Isle of Man (1973) 51 people died and 8 seriously injured. In both cases the materials individually complied with then current requirements, but the way they had been used and the design and use of the buildings were the issues. We will be returning to fire in other parts of this book.

To recap:
The design team could comprise:

- architect contracted to the client
  - converts the clients ideas into formats that others can work to – mainly graphical and written specification
  - obtains legal approval for the proposed work – Planning Permission and Building Regulations
- principal designer of the project
  - appoints consultants for their specialisms – other design team members
  - oversees the project during construction
  - signs the completion certificate to certify the work is to the standard as set out in the contract documents.

Chapter 3

The construction team

In order to convert the clients' scheme into reality the design team has produced working drawings, specifications, obtained the various permissions and costed the whole project. Now a 'builder' will be required.

Depending upon the size and complexity of the project the builder may be one person or an army of construction workers known as the construction team. This chapter provides an overview to the evolution of the different trades involved in construction. Simple historical accounts show how and why the trades have evolved and why true skilled tradesmen (that includes the many females in the trades) are not inferior to the professions. Different skills are needed to complete the clients' projects; some are practical whilst others are theory-based; they all had to be learned.

As the client employed the architect to head the design team, so the 'builder' heads the team to construct the project. Builders can be individuals who have one particular trade but will carry out most aspects of building. They are general builders and employ others to help as and when needed. There are also tradesmen who specialise in one particular aspect of building (see Figure 3.1). Over the centuries the range of trades has increased as different building techniques become popular, but the main core are:

- masons
- carpenters
- plasterers
- plumbers
- electricians
- glaziers
- groundworkers
- painters
- roof tilers
- scaffolders
- steel fixer.

Starting in the order of age, the mason and the carpenter must be the earliest trades.

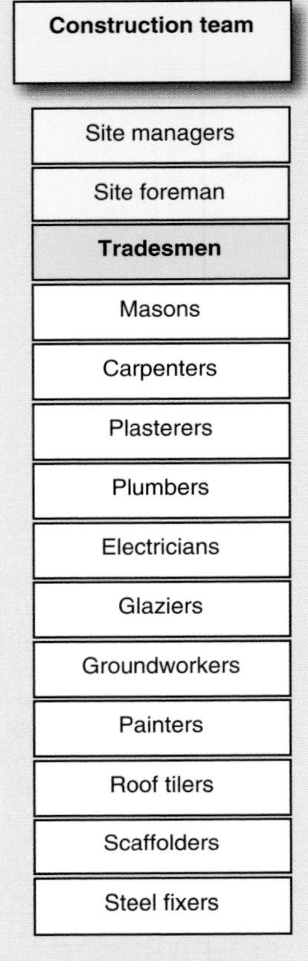

Figure 3.1 The construction team.

3.1 Trades

Masons

Masons are people who work with stone, burnt clay bricks, concrete blocks etc. Most commonly one of the early trades on site who build brickwork walls and chimneys are referred to as bricklayers. There are various specialisms within the trade of masonry. Hewers or cutters cut, work and dress (cut to shape) large pieces of stone, and those who set (lay) the stone are all stone masons.

Historically local stone to the site has been used due to the problems of transport. Exceptions would be where clients want to display their wealth or religious buildings where wealthy benefactors would make very large donations. Commonly sedimentary rocks such as sandstone and limestone have been used to build walls (see Figure 3.2). In London, for example, there are no natural rocks near the surface therefore stone had to be brought in by barge from near Maidstone in Kent to build the Roman city walls. They built two outer skins of dressed limestone 4.50 m apart and filled the cavity with ragstone (limestone) built as quarried (rubble wall)

Figure 3.2 A Georgian house in limestone.

i.e. not dressed (see Figures 3.3 and 3.4). Unlike dry stone walls of Northern and West of England, 'rubble' walls were bedded in mortar (clay and lime or lime and sand).

For important buildings and gatehouses, larger stones were cut as blocks in the quarries and sent to the site, where the master mason would select suitable stone blocks to be shaped based on the grain, colour and texture of the stone. Stone would be brought hundreds or even thousands of miles to achieve the desired finished result. Marble used for the stone temple to Claudius was brought by ship to Colchester from the Purbeck Hills near Poole in Dorset. Several centuries later the Normans brought limestone from Caen in Northern France. The particularly fine grain termed a freestone has no apparent bed (layering effect like leaves of a book). Therefore the stone could be cut in any direction, making it ideal for the tracery found in Norman windows (see Figure 3.5). Freestones were, and still are, very valuable, therefore the masons who worked with the stone were known as freemasons

Figure 3.3 Part of the London Wall.

and master masons of their guild. (Guilds were people who carried out the same type of job who organised themselves into groups/guilds to protect and further their knowledge. Guilds even today are very influential in the running of the country's affairs.)

During the Stuart period after the Great Fire of London in 1666 only a few of the stone buildings survived. Most buildings in the fire area had been built in timber frame, leaving a pile of ashes. New laws were passed and masons rebuilt the city (see also Chapter 7).

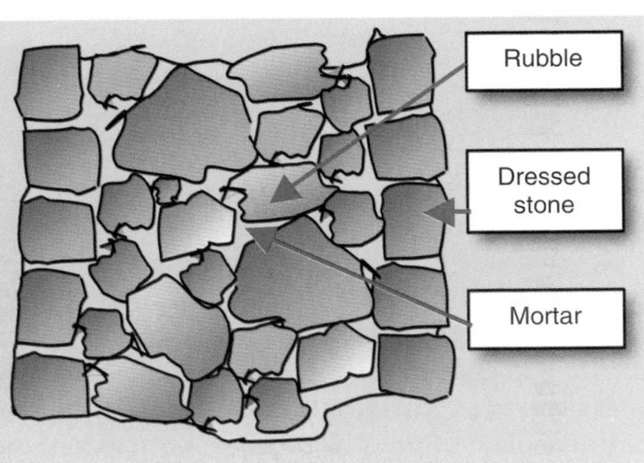

Figure 3.4 Section through a rubble wall.

Nothing much has changed over the centuries. Important buildings still have stone walls and floors, albeit more commonly veneers of stone attached to concrete or steel frames. Tower Bridge, for example, (completed in 1894) is an early example of steel frame clad in stone (see Figure 3.6). Canada Tower in Canary Wharf London is a modern steel-framed building with the foyer clad in over 8000 m^2 of thin marble slabs to the walls and floor.

Dressing stone is very expensive both in terms of time taken and the method of quarrying. Areas such as Bath are famous for the limestone buildings, some dating back over two thousand years. However, the stonework has been dressed into blocks and bonded

Figure 3.5 Gothic style tracery.

for strength and stability. Centuries later the Saxons took over the building programme in England and many of the skills of the mason were lost in favour of the trade of carpentry.

Clay in the south of England gave rise to brick manufacture. Cheaper than stone, more locally sourced and easier to use, the masons became bricklayers. Sorry if this is becoming tedious, but once again the Romans brought the skills of brick making and laying two millennia ago. They had conquered the Belgae tribe who had mastered the art of burning bricks. With the skills of brick burning the Romans set-up some of the earliest brickworks using the clay around north Essex. Brick buildings indicated wealth and importance. Ashlar (large stone building blocks) buildings displayed even greater wealth and prosperity. Stately homes, large manor houses, churches in prosperous areas and cathedrals often have ashlar walls (see Figure 3.7).

In the 1920s small firms produced slabs from the crushed clinker termed 'breeze' left over from commercial coal-burning boilers. They mixed the

Figure 3.6 An early example of steel framed building: Tower Bridge.

breeze with Ordinary Portland Cement (OPC) and placed it in moulds to form large blocks commonly known as 'breeze blocks'. As an apprentice I met old bricklayer who could remember the introduction of breeze blocks. They were sold as cheap alternatives to bricks. Covering the wall area of six bricks with one block, they were faster to lay, although they were very heavy. There was no regulatory guidance as to the crushing strength at first therefore they could only be used as infill non-loadbearing partitions. The 'Breeze block' company ceased trading in 1952 when aerated autoclaved concrete (aac) was introduced from Sweden. Today aac is now known as aircrete. There are many different types of concrete blocks, from clinker through to aggregate. However, just as with Hoover and vacuum cleaners, some people still call grey blocks 'breeze blocks'.

Carpenters

It could be argued that carpentry is the oldest trade, but that can be left to others to debate. Carpentry is primarily working with timber. Timber is a felled and worked natural material from a tree, mainly the trunk, though boughs (largest of the branches) are also used. In southern England carpenters could use the hardwoods English oak, elm, ash and chestnut. The first three can live for hundreds of years, becoming wide around the girth (measurement around the trunk of the tree) and providing the large baulks used for posts and beams.

Like masons, carpenters would diversify into several different groups. After the woodsman felled the trees the main trunk would be stripped of the

Figure 3.7 A fine example of a Georgian mansion with ashlars.

bark and squared up using an adze (similar to an axe but where the head is at right angles to the shaft). Later the large double-handled pit saw was developed, enabling the sawyer to cut baulks into planks of timber. The sawyer and his mate would work in a pit, between them sawing the length of the log (felled/cut down tree trunk) whilst the timber was still green. In construction the term 'green state' indicates the material contains water that will eventually dry out. Concrete, bricks, timber etc. all go through a 'green state'. Oak, for instance, is more easily worked whilst in its green state. When fully seasoned and with a moisture content less than 20%, the wood becomes very hard.

Although during the Roman occupation (43–410AD) they had all of the skills to produce fine furniture, ornate windows and doors, when the Romans left Britain most of the skills were lost and had to be re-discovered.

Invaded by Angles, Jutes, Vikings and Saxons (410–1066AD) Britain lost many of the skills it had acquired from the Roman occupation. All new buildings once again returned to local materials, mainly rough worked timber, reeds and straw. Although in the latter years of Saxon rule masonry and carpentry skills became more refined in prestigious and religious buildings.

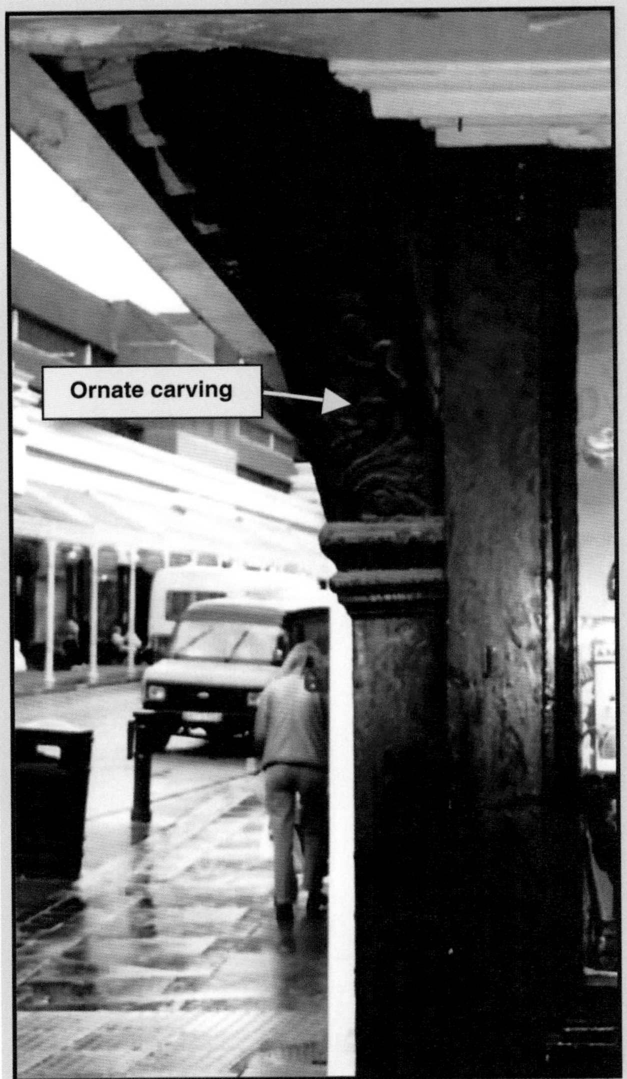

Figure 3.8 Ornate Jacobean carvings.

With the Norman conquest (1066 AD) came greater skills for the mason and carpenter. Later, during the Tudor period (1485–1603) the carpenter became the main builder, with the development of the timber-framed houses and the introduction of the furniture maker. During the Jacobean period (1603–1660) more ornate carvings were added and fashion could be seen in the building fabrics (see Figure 3.8). As the carpenter's tools developed, the trade of cabinet maker was born (1660–1720). The previously little-used walnut tree became prominent and gave its name to the period as the fashionable wood for furniture making.

Excellent examples of the skills of the carpenter can be seen in the Victoria & Albert Museum in London, where furniture, pattern books, ornate panelled falsework and linings can be seen. To plan your visit, log on to www.vam.ac.uk/collections/british_galls/galleries/54/index.html.

So far we have looked at the trades of the mason and the carpenter. Both have been the main builders throughout history. Secondary trades such as tilers, plasterers, blacksmiths and leadworkers have enabled magnificent buildings and monuments to survive to this day as working structures.

Plasterers

Plastering was originally carried out by women who spread a mixture of clay, slaked lime, straw, dung, and sand over the rough surface of masonry walls and the basket woven sticks termed 'wattles'. The process was to daub the mixture, pushing it into the small gaps between the stone or 'wattles' – hence 'wattle and daub'. Later in Tudor times animal hair and bull's blood were added to provide a pink colour and fibre to reduce shrinkage and prevent the dried daub falling off the wall. At the end of the Jacobean period ornate decoration on buildings was fashionable. Those who could afford carvings and moulded bricks displayed their wealth. Coloured bricks were introduced to form patterns in the walls such as Tudor checker boarding (see Figure 3.9).

Those who could not afford masonry had their timber framed buildings plastered and lined to appear like ashlar/stone blocks (see Figure 3.10). The more wealthy would have moulded patterns known as pargetting or freehand sculpture onto the plaster (see Figure 3.11). Later plastering became fashionable inside buildings. Again it was only the wealthy that could afford plastered walls. Churches and cathedrals were commonly flat plastered with bright painted scenes of religious happenings – the start of the painter and decorator (see Figure 3.12). Modern day plasters tend to be gypsum-based powder mixed with water and trowelled onto the walls or ceilings.

Plasterboard introduced from North America in the 1940s became popular for ceilings taking over from the standard lathed and plastered predecessor. The trade of tackers had evolved. Tackers nail or more commonly screw sheets of plasterboard onto the underside of ceiling or upper floor joists prior to the plasterers skim coating with gypsum plaster. In the late 1960s plasterboard was used for dry lining on timber or metal studwork for partitions and timber frame buildings. In the 1970s plasterboard was used as a dry finish fixed to masonry walls using dabs or battens. Another trade of dri-liners or dri-walling contractors has grown, where plasterboard is screwed to studs and the joints filled with jointing compound. The process is virtually complete and ready to paint or wallcover in a few hours (see also Cooke (2007), pp 303–8).

Figure 3.9 Examples of Tudor decorative brick patterns.

Plumbers

Although prestigious buildings and those of the rich had plumbing way back in the 14th century (yes the Romans had plumbing 14 centuries earlier), the main bulk of dwellings would not have had any internal water or sanitation until the Victorian period. The word 'plumbing' comes from the Latin 'plumbum' or 'plumbus' meaning lead. For many years lead had been the main metal used for pipework, taking over from wooden pipes and conduits. Valves and cocks (original word for taps/faucets) were formed in brass as it is easier to cast into intricate shapes and more durable than the soft metal lead.

As the materials have changed over the centuries plumbers now work with an array of metals and alloys such as copper, stainless steel, lead, brass, zinc plus several types of plastics such as PVC, polythene, ABS, CAB, acrylic and polystyrene (see Figure 3.13). Not only working on drinking water supply the plumber will be involved with waste water, sanitation and drainage. Specialist

Figure 3.10 Stuccoed mock ashlars.

plumbers work with sheet cladding metals such as lead, copper and zinc covering roofs, mansard walls, guttering, and flashings. Many historic buildings require maintenance, therefore the skills of the plumber/sheet metal worker need to be kept alive.

As 'town gas' took over from oil lamps and candles for the lighting of buildings and eventually heating buildings, the plumber became the gas fitter. Later in the 19th century came the introduction of electricity as the newer, cleaner and silent fuel for light in buildings and the birth of the electrician.

Electricians

With modern buildings electrical services can account for a major part of the building's cost. Not only do electricians install cabling, switches and sockets to allow access to electricity anywhere in the building, they are also involved in installing security, safety, and communications equipment.

Glaziers

The earliest known glass window was found in Kent and dates back to around 200AD (yes, the Romans again). It took about another 900 years before glass

Figure 3.11 Very ornate pargetting.

for glazing became fashionable. Small panes were cut from Norman slabs and held in lead cames (cames are small 'H'-section strips that can be soldered together to form a lattice of slots to hold small pieces of glass) (see Figure 3.14). Only the wealthy (mainly the Church and aristocracy) could afford the lead lights (lights as the name suggests are parts of a building where light can naturally enter). Unfortunately there is not enough space in this book to cover the history of glass. For glazing therefore we will have to start with Sir Alistair

Figure 3.12 New hand made copies of original Regency plaster cornices with enrichments.

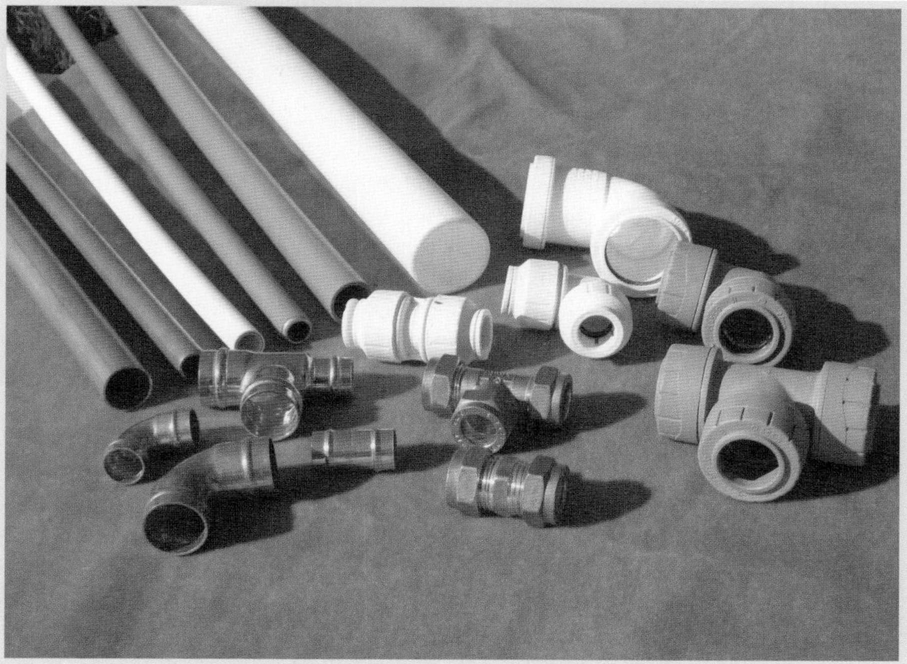

Figure 3.13 Plumbing pipes and fittings.

Figure 3.14 Leaded light casement windows (not original).

Pilkington's float glass. Most of the glass in modern buildings is float glass. Optically perfect, the super-cooled liquid can be cut into small panes or very large pieces termed 'plates' to form complete glass walls (you may want to also read Cooke (2007), pp 340–43).

Glaziers cut, work and install glass into many different materials: wooden windows, leaded lights, copper lights (old-fashioned fire-resistant glazing before wired glass or pyroglass), metal windows and directly into stone, brick or concrete openings. Glass can be held in place with putty, tape, washleather, felt, polysulphide, silicone, butyl or acrylic mastics or in neoprene plastic glazing gaskets. Glaziers cut holes in glass, smooth the edges, brilliant cut, acid etch, or sandblast the glass. Handling very large and heavy glass units requires specialist equipment (see Figures 3.15 and 10.18).

Groundworkers

Formerly known as labourers, the skills of ground working are now recognised. Digging trenches by hand is both heavy and relatively unskilled work.

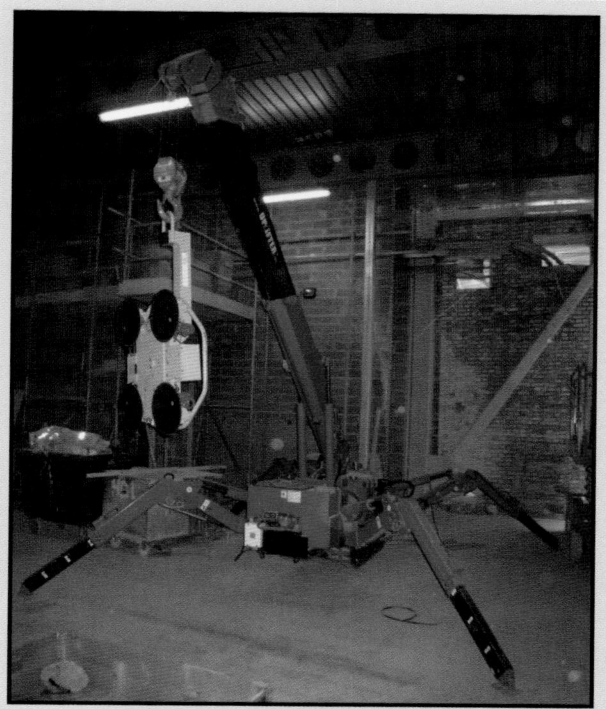

Figure 3.15 Spider crane with sucker attachment.

Thousands of men came over from Ireland looking for work in the 18th century during one of several famine periods. Being tough and strong, they dug the canals and sewers by hand. They were employed as 'navigators' as they cut through all terrains under the supervision of the great engineers of the day. Abbreviated to 'navvy', the term became attached to men who would dig anywhere. As the work of digging canals and sewers dried up they turned their skills to tunnelling. Most other 'navvies' found work on road building and labouring on building sites digging trenches for foundations and drainage runs. If you have ever tried digging a trench you will know it requires strength and skill to produce vertical straight sides with a clean formation (bottom of the trench).

Modern groundworkers on many occasions still dig trenches using hand tools like spades, grafts, picks and shovels where machinery cannot be used. They have to provide trench support such as timbering or sheet piling to prevent the sides collapsing. They remove ground water when working below the water table, freeze fibrous ground, and work to given levels. They place concrete for foundations, ground floor slabs, lay drainage runs, build manholes and inspection chambers and other service utilities such as water, gas, electricity, and communication equipment (see Figure 3.16). They place steel reinforcement as bars or cages prior to pouring concrete anything below dpc level. Today there is the trade of the groundworker.

Figure 3.16 Exposed underground utilities.

Painters

Painters also include decorators. The art dates back as long as man himself decorating the caves and the structures around him. Modern painters, however, have to prepare surfaces prior to coating them with protective film mainly derived from plastic and metals. Paints such as acrylic and emulsions are basically plastic adhesives that glue pigments onto surfaces. Emulsion paints are mostly PVA (polyvinyl acetate) adhesive emulsified in water with coloured pigment that when dry forms a plastic coat over the material. Plaster walls are porous unless polished, therefore emulsion paint both seals the surface and can provide colour to the wall. Most materials require a paint film that adheres to the surface after the carrier (water or spirit) vapours off known as a 'primer coat'. Metal surfaces such as radiators and pipes require special paints known as 'metal primers'.

Undercoat paints are then applied to provide thickness to the film and, having large quantities of pigment, they provide good opacity (opacity is the density of colour that prevents vision through the paint). Finally, finishing coats should be applied, providing the desired effect, be it high gloss, silk or matt finish. If used externally, the finishing coats should have ultraviolet stabilisers to prevent yellowing and discoloration from sunlight and prevent the loss of elasticity. The finishing coats have to be weather-resistant, and to an extent resistant to abrasion.

Painters can disguise materials to look more expensive. For example, marbling and graining over wood or plaster is achieved by skilful paint techniques (see Figure 3.17).

Figure 3.17 Peter Woods demonstrating the use of a 'flogger'.

Sheet materials have been used to cover walls for over 500 years. Early examples of wallpaper from the 17th century (Georgian period) can be seen in the Hollies Museum in Colchester and in the Victoria & Albert Museum in London. Painters and decorators gild precious metals over wood such as gold leaf, they sign write shop signs, pub signs and the carvings, decorate the highly ornate plasterwork of cornices and covings (see Figure 3.12).

Roof tilers

Clay tiles have been used in the Mediterranean countries for thousands of years. Pegged low pitched half round tiles provided excellent weather resistance and a degree of fire resistance. Smaller plain tiles became popular with the Norman and Saxon buildings. There are many fine examples of original clay tiled roof in Kent and Essex including the Norman castle that currently houses Colchester Museum. The 'Old House' in Rochford, Essex is a fine example of a 13th century dwelling with a clay tile roof (see Figure 3.18). The

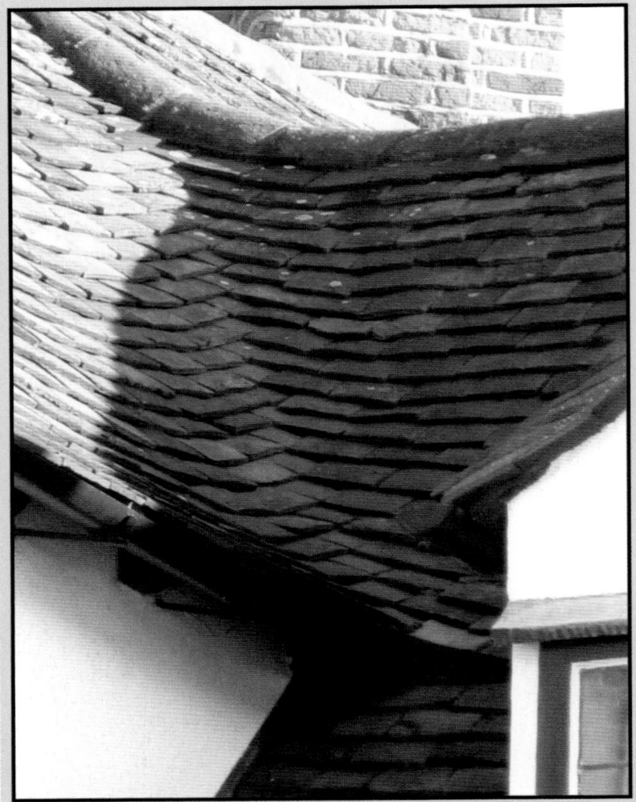

Figure 3.18 A herringbone valley in hand made tiles.

original roof may have been thatched with local reed however the tiled roof had been there for several centuries. On a recent renovation a retired roof tiler had to supervise the valley tiling in the traditional method as the skills had been almost lost (look closely how the valley has been finished).

Slate roof coverings took over from clay tiles mainly with the advent of the railways in the 19th century. Some wealthy building owners had slate roof coverings brought around the coast line from the West Country and Wales. In the Victorian period slate roofs were the choice of most designers. Slate from the Lake District is thicker and slightly more coarse in texture than the Welsh and Cornish slates. The last working slate mine in the Lake District is the Honister mine. It is definitely worth a visit to see the slate mined and worked with much of the operation still by hand crafting (www.honister-slate-mine.co.uk).

Scaffolders

To access the work at all heights above 1.5 m, scaffolding is normally used. With recent changes to the safety requirements, only trained and qualified

Figure 3.19 Examples of tied scaffolds.

people can erect, dismantle or adjust scaffolding. Scaffolding can be a series of metal tubes held together with couplings as a tied scaffold or independent scaffold (see Figure 3.19). Scaffolding can be used to protect areas or people, enable safe working at heights on tower scaffolds. GRP (glass reinforced plastics) scaffolding and access towers are available where electric-sensitive areas require access.

Scaffolding on large structures has to be designed prior to assembly and requires skill and experience (and a lot of nerve even to erect it on very tall buildings). Scaffolding is a trade in its own right.

Steel fixers

Since the end of the 19th century the trade of steel fixer has been associated with buildings, for example Tower Bridge (1894) and The Ritz Hotel (1905). Steel fixing no longer relies on hot rivets but modern buildings have nuts and bolts or welds.

So far we have looked at the background of most of the important trades and their offshoots. They all require lengthy training and experience, plus qualification. It has been a fact that through the centuries of building there has always been an 'us and them' mentality. The professions, thinking they were superior as they kept themselves clean, tended to have higher qualifications and generally earned more prestige and money. Look back at some of the magnificent structures through history. Who is remembered? Was it the riveters that assembled the plates that made the bridges over the River Seven? Was it the casters who cast the metal latticework for the Ironbridge in Shropshire, the stone masons who built the stone bridge piers that hold the Bristol Suspension bridge? No, we acknowledge the designers and engineers like Telford, Derby, and Brunel. Today we have some of the most spectacular structures in the world designed by teams of people, yet it is the leaders who are remembered.

3.2 Why does history remember the designers?

Without the design, imagination and often sheer risk takers, these exceptional structures would never have been built. The 17th century remembers Wren, Inigo Jones, Gibbons, Nash and Adams. Today in the 21st century we have Lord Norman Foster and Lord Richard Rogers, who are arguably two of the most world renowned architects. However, there are many fantastic buildings designed by other architectural partnerships such as One Canada Square, the tallest office block in Europe, designed by Pelli Clarke Pelli Architects of Connecticut USA, Renzo Piano, architect of The Shard, which when completed in 2009 will be the tallest office block in Europe standing at 308 m, and Skidmore, Owings & Merrill – SOM, the architects who designed 201 Bishopsgate and the Broadgate Tower, the third tallest in the City of London.

The buildings, though, are erected and finished by the armies of tradesmen. Like all armies, direction is needed and communication. It would be impractical for the designer to direct the builders. Wren tried it and history writes he often had disagreement with the employers/client and the workforce of tradesmen.

Today major building contracts are 'management team' led (see Figure 3.20). The main contractor assembles a team who oversee the whole building project (see Chapter 10).

3.3 The on-site team

The team would be headed up with a project director. Responsible to the main board of directors, he or she would make executive and policy decisions to complete the project on time, within budget and safely. Clients of large developments often set aside quite sizeable sums of money to pay bonuses to the

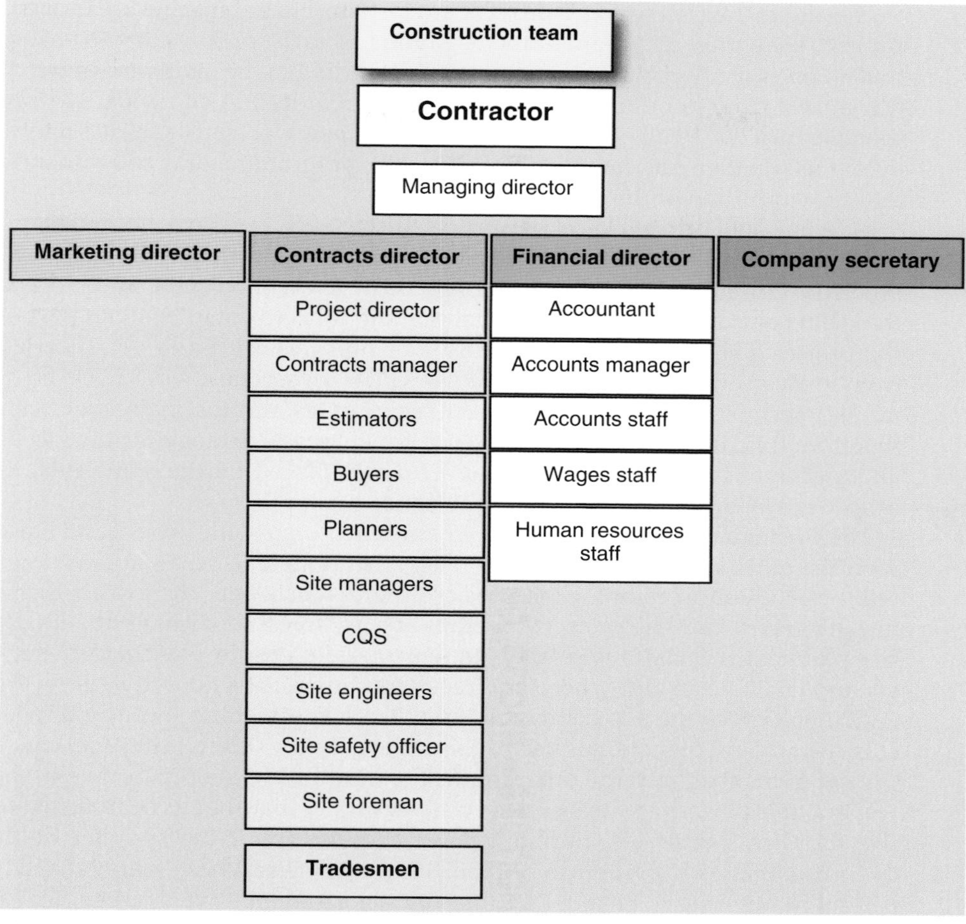

Figure 3.20 Management led team.

workforce to work safely. Accidents, apart from the suffering of the victims and their families, cost the contractor lost work hours, extra replacement of labour and materials, and the client time before they can use the building.

Site managers and project co-ordinators

Safety is improving on the larger sites by appointing specific people to look out for bad working practices and foreseeable hazards. Foremen or team leaders (sometimes referred to as project co-ordinators) are responsible for sections of the project and communicate between the site managers and the trade foremen. The site managers will work to the project plan and amend as appropriate. The foreman then hand out the building tasks to the appropriate trade foremen, gangers or chargehands (different trades have different names for the person who leads the tradesmen).

The site managers on large projects are responsible for specific areas of the project. They meet as a team at the beginning of each working day to ensure that all parts of the project are progressing according to the main plan and that communal plant and facilities are available as required. A site with, say, five tower cranes will need efficient preplanning of their work programmes. However, the weather can completely change the programme and consequently affect the duration of the project.

Materials will be needed. If they are on site too early they may become damaged, they may be stolen, they may be in the way and they will need to be paid for. Therefore the tradesmen, many of which will be sub-contractors to the main contract may only supply labour and tools to complete their part of the project. The materials may be heavy or bulky and need to be offloaded from lorries and placed in the area of work. The tower crane would be a communal piece of plant to be used by sub-contractors and the main contractor therefore time has to be booked. The site foreman will arrange for time to be allocated and if necessary specific attachments placed on the crane such as brick clamps, lifting webs, chains, shackles or hooks.

The site managers will organise their particular part of the project and liaise with the other site managers as previously mentioned. Regular site meetings will be required to monitor the project progression where the project director will ensure that the project is progressing on time and within budget. The site planners will liaise with the site managers to ensure plant and materials are available as and when required. The contractor's quantity surveyors (CQS) will check the sub-contractor's day work sheets, time sheets and measure the amount of materials used against the delivery tickets. They will check the use of plant against the programme. For example, waste removal vehicles, waste skips, delivery pallets and special packaging that should be returned to the supplier. Pallets etc. normally have a deposit that the supplier holds against damage or non-return of their delivery pallets. Waste removal needs careful monitoring to ensure that the contractors are not overcharged. Skips, for example, can be difficult to account for as there is not a measurable material being transported.

Planners

The planners, CQSs, site managers, foreman and sub-contractors will all monitor the call-off of materials. It used to be easy to add a few bags of cement, sand and bricks to the order for 'private' work. The builder or sub-contractor would visit the local builder's merchant yard and verbally add whatever they wanted for their own purposes. When the merchant tried to add the extra material to the contractor's invoice there would be dispute. Today, large contracts have all of the materials and suppliers arranged via the company buyers. The site can then either order materials against a call-off programme quoting the order number or place the requisite order through to the buyers to order and supply the materials. Theft still happens, though, but generally it has to be well organised.

Buyers

Buyers will generally liaise with the estimators to obtain the best price and service from the suppliers. The estimators would have based their predicted costings on established prices or quotes with suppliers at the pre-tender stage. The paperwork would then be passed to the buyers to arrange for the supply to take place.

Buyers via the accounts department will apply for credit accounts with suppliers who are likely to obtain and supply materials and plant over the duration of the contract. Credit accounts are generally monthly however I have known some very large contractors to negotiate extended credit facilities. Credit accounts enable materials and plant to be delivered to a contractor's works or delivered to site and invoiced for payment every 28 days. If the contractor pays in full a discount – commonly 2½% – is given. If the payment is late then the discount is foregone. If the payment is very late additional interest may be charged. However, in real life the supplier/merchant has to compare the possible loss of future orders against the interest incurred by late payment. The merchant or supplier would normally have a credit account with the manufacturer or factor or agent. They, too, provide discount for early payment. They may also offer additional discounts or rebates to merchants that sell very large volumes of their product, therefore the merchant may use the extra discounts to offset the loss incurred by late payment from the contractor. The manufacturer has raw material suppliers who have similar trade credit agreements and thus there is a chain of credit actions and reactions. The main method to try and overcome the problem is to limit the amount of credit and then stop the supply of future orders, known as 'blacklisting'.

Accounts department

Often treated as the 'ugly sister' of the company, this is perhaps one of the most important departments. They will make the arrangements to open the credit account on behalf of the contractor. To open an account most merchants or suppliers use a credit reference agency to gauge the amount of credit per month that can be given to a trading company. The agency will provide a credit limit based on credit-worthiness of a trader and details of any court actions needed to recover debts. The accounts department will also have to give details of the bank the company uses.

Where a contractor does not have a credit account with a supplier or manufacturer a pro-forma would normally be raised. The supplier invoices the contractor, who will pass it to the buyer to check the quantity and price against the order. The buyer then passes it to the CQS to check it against the tender documents and they send it to the accounts department to raise a cheque to pay for the materials before they are delivered. There is no credit arrangement therefore the cost is covered by the contractor from the outset.

Materials and goods that have been supplied to the site, collected or delivered to the contractor's works will be invoiced to the company. The invoice

Figure 3.21 Delivery vehicle with mobile forklift attached.

will be checked against the orders and job number. (Job or contract numbers are very important where the contractor may have several contracts running at the same time all with call-offs on the same supplier/merchant.) The CQS will have to check the orders against the tender documents and the main order set-up via the buyer. Delivery tickets should have been gathered under the supplier's name, or date delivered.

Delivery tickets are very important documents. Suppliers usually supply the delivery driver with two self-carboning tickets showing the date, delivery address, goods and or materials, any pallets or containers that should be returned and have a deposit on them and any services provided. For example, in the early 1970s when a brick lorry arrived on site most of the trades had to help unload and stack the material, termed 'handballing'. The driver would stand on the back of the lorry and throw either three or four bricks at a time and we had to catch them and stack them on the site. Timber was delivered on flat backed lorries a 'standard' at a time, all having to be offloaded and stacked by hand. Everything other than sand and aggregates was 'handballed', as the driver's liability ended at the tailboard.

Today in the 21st century lorries have Hiab cranes with brick clamps, straps and hooks or small fork lift trucks carried on the back of the lorry so the driver can self-offload (see Figures 3.21 and 6.3). The service is chargeable, and therefore shown on the delivery ticket. The most important details, though, are the official order number and the signature of the person taking the delivery. The two carbon signed tickets are separated and one left with the site and the other returned to the supplier. We will be looking at the whole process in full in later chapters.

Chapter 4

Stages of design

Following on from the design team, this chapter outlines the various stages of the design process, drawing analogies with everyday decisions such as buying a new coat or car. Using a simple scenario, the stages and reasoning for each stage will present a trial from inception through to tender stage.

National Governments orchestrate the development of the Built Environment by directives and Acts of Parliament. Policies are made by processes of lobbying, financial influences and reactive decisions, most of which are out of the scope of this book. To conclude the chapter, two topics for group discussion have been suggested.

Politics are seen by many as boring. However, like it or not, it controls the way we live. Being a politician is not an easy occupation. What may appear simple to some may have many sub-influences, even foreign, that prevent progression. Try debating the topics as groups; it is far from easy.

There are four main stages of design:

1 Inception – the idea originating from a want or need of the client.
2 Feasibility – whether the idea is possible, financially, legally, practically.
3 Design – converting the idea into a tangible plan that others can read and work to.
4 Procurement – obtaining the materials, workforce and equipment required to produce the idea.

Wanting something like a new coat or car or even somewhere to live could be classed as 'having an idea'. Some people then go out and look for shops that sell the item. Others may prefer to look at a catalogue, magazines or perhaps trawl the Internet in hope they will find what they are looking for. What though if you want something different from the goods that are available? You may be a different shape or size from the 'standard' person the item was made for therefore can it be modified to suite your needs? Do you ask other people what they think about what you want? What happens if you see exactly what you want but it is far out of your financial reach?

Let's look deeper at the issues of trying to buy a new coat. Your employer is sending you on an outward bound course in Scotland in February. What do you know about the special requirements of the coat?

Should the fabric be:

- well insulated
- waterproof
- windproof
- have ventilation

or would it be better to have a really fashionable jacket because you don't want to feel a 'wally' in the evenings?

As you can see, it can be difficult enough to buy something a simple as a new coat. At least there are coats you can try on and actually see the finished item. Now compare with buying new business premises.

Example

You are a manager in an insurance company. Your staff is located in different small offices in several buildings in a small town. Most of your business is based overseas therefore you rely on the telephone and computers. Business is good and increasing and therefore a decision has to be made whether to rent or buy more premises, to rent or buy another larger premise and move all of the workforce into the one building, or to have a new building specifically built for your company's needs.

As a manager you are probably an employee therefore the owners of the company will make the actual decision.

The company directors agree that a new building would be a good idea. Perhaps something really different like the 'Gherkin' 30 St Mary Axe in London (see Figure 4.1). Not many people out of the world of insurance had heard of 'Swiss Re' before the Gherkin was built. Having a spectacular building will act as an advert as well as providing an image of a very successful company.

Figure 4.1 No. 30 St Mary Axe: The 'Gherkin'.

How much is it going to cost? Where can the new office be located? There are many questions to resolve.

Location may be relatively easy to decide. The directors do not want to move the company out of the area as it will require moving all of the staff. It can be very costly financially and possibly losing some of the staff would not want to move home.

Is the next step going to be looking at all of the real estate agents and land agents for a plot of land? Alternatively, would a project management consultant be a better option? If the company has built up a good reputation in the field of insurance it is unlikely they will have the professional know-how and time required for finding suitable land.

There are several methods of meeting the client's wants and needs. The traditional method of employing an architect has been replaced with the term 'lead consultant'. It may still be an architect who leads the design team, but another professional such as a surveyor could take the position. In either case the design team would be lead by one person named in the contract between the client and the lead consultant.

4.1 Choosing an architect

Having a new building designed and built could be likened to buying a tailor-made coat; in both cases the client has a 'want' but cannot see the finished item until the design, planning, and construction has taken place. The client can gather ideas of the type and style of building they would like by visiting other companies, other towns and cities and even other countries. If a building design or style appeals enquire the name of the architect or designer. Many commercial buildings have a plaque with the name of the designer on them. All new buildings have a building manual stating how the building should be maintained. The name and addresses of the designer and the builder would also be included. An alternative would be to contact the RIBA (Royal Institute of British Architects), who have lists of architects and architectural practices that are experienced in specific projects. Other professional bodies include the Royal Institution of Chartered Surveyors (RICS), and Chartered Institute of Architectural Technologists (CIAT), and the Institution of Civil Engineers (ICE).

The selection process

The client may have contacted several designers therefore a selection process will be required. From the consultant's perspective does the client want something that is so far from the style of their business? The initial interview will influence the decision whether to enter into a contract. There is no hard and fast rule regarding the initial meeting. It may take place at the client's premises to enable the consultant to obtain a feeling or mood of the client's business. The image of the company will be influenced by the new premises. For example would you want to buy your shopping from a tatty old house where all of the stock has been piled up in various rooms? The items may be in date and in perfect condition, yet as a consumer you expect a well organised shop, brightly lit, items in order and on display and a pleasant atmosphere to shop in. If the client intends to do business from the new premises, image is important. Old vehicles parked outside an ugly building does nothing for the image.

Alternatively the client may wish to have the initial meeting at the offices/practice of the lead consultant. In a similar way to establishing a feeling the client will be less than impressed being welcomed to a dingy suite of rooms where space has to be made before the client can even sit down. (There are architects who produce excellent designs yet work in awful environments. In essence, is the client buying the service or being put off by the experience?)

The client should attend the offices of the lead consultant at some stage early in the selection process. For example, Sir John Soane was a famous architect in the early 19th century. Son of a bricklayer, he had a practical upbringing and an eye for buildings. He trained for many years and became perhaps one of the most famous of the Regency architects. For much of his working life he was involved with the design and overseeing of the building works of The Bank of England in Threadneedle Street, London. However, all that remains now of his work is the high windowless outer curtain walling.

During his training he stayed in Rome and gathered various marble sculptures and parts of buildings, which he had shipped back to England. In 1792 he bought number 12 Lincoln's Inn Field in London as his family home and based his practice in the rooms at the back of the house. He later bought number 13, the house next door, which he demolished and re-built as his new home. The rear of the building once again became offices for his practise, including speciality rooms to house his collections of marbles, fine art and picture galleries. He used parts of his home when entertaining clients, displaying some of his inspirational ideas of design and construction; for example, the use of mirrors strategically set at ceiling level and between bookcases to create an impression of continuing space and to reflect borrowed light.

Soane incorporated loggias onto the frontage of his new home. Loggias are of Italian design, and had a renaissance in Rome in the late 1800s and were therefore very fashionable, especially in London in the early 1800s. (They are similar to a balcony with a roof above. The front may be open, with or without a small wall defining the front edge. The ends may or may not be enclosed. They were originally designed for wealthy homeowners as a place to sit in the open air and be seen, but they could also be used as an external corridor between rooms in larger buildings.) The climate in London is cooler than in Rome, so at a later stage Soane redesigned them by enclosing the fronts with vertical sliding sashes, thus extending the area into the room. He used large panes of plate glass to glaze the sash windows; the most expensive of glazing glasses, such that not until 1845 when the duty on plate glass fell by three quarters did it become more popular.

Another innovation by Soane was to design horizontally sliding shutters glazed in silvered glass that ran parallel with the windows, housed within the thickness of a false façade. The effect when in the closed position formed framed mirrors facing back into the room, creating the illusion of greater space and reflected light. Domestic lighting in Victorian times was both expensive and of poor quality, mostly from candles or oil lamps. Some domestic buildings had coal gas lamps comprising a flame within a decorative glass bowl. However, like candles and oil lamps, the available light was still poor. The gas mantle was invented in 1886, 49 years after Soane's death. It was commonplace in the better class of buildings of the wealthy to have panelled wooden window shutters serving the function of added security and when not in use concertinaed back into the reveals of the wall.

His home and office are open to the public and remain virtually unchanged since the day he died. For further information about the enormous range of buildings that Soane designed, look on the following website: www.soane.org/soanebuildings.html. If you are able to visit London, why not look around the Soane Museum? It is about a 5 minute walk from Holborn underground station and has free entry; for groups of visitors a short lecture can be provided for a small fee. Their website is www.soane.org. When you visit you will see his infamous floating domed roof that even today modern architects still try to copy. It is well worth a visit.

Although architects of today perhaps may not have such a display of antiquities as Soane, they will have samples and scaled models of their previous work. Electronic displays and computerised virtual reality enable their clients an insight into the innovations and skills that can be offered (see the description of Arup Associates in Chapter 10).

Considering an architect as the lead consultant

The RIBA have produced a document, the 'Standard Agreement for the Appointment of a Consultant: 2007'. Within the document is a list showing what the client should expect when appointing the lead consultant headed the 'Standard Condition of Appointment for a consultant'. They can be outlined:

- The brief: The basics of what the client wants. This should include the objectives and use of the development. For example, it is not sufficient to say, 'I want a new office block to be built in Swindon.' The lead consultant will ask or expect to be provided with detail of how the client intends to use the building and for how long. How many people will be working in the new building? Will the number change significantly? Is expansion of business likely? How many hours a day and days of the year will the building be used? The brief could be considered an overall plan.
- Construction costs: At the early stage of appointment it is unlikely for the lead consultant to provide anything other than an approximate costing, called an 'estimate'. The approximate costing is a guide price only and should be accepted as within about 25% error of the final price. There are so many unknowns at that stage it would be impossible to expect any more accuracy.

After the initial meeting and the selection of the architect a contract will be produced.

In 1968 the RIBA produced a set of guidelines known as the RIBA Outline Plan of Work. The documents have been updated as required. In the Plan of Works 1998 there were three main divisions or stages:

1. feasibility stage
2. preconstruction period
3. construction period.

4.2 Feasibility stage

The first stage, as the title suggests, is concerned with the question: 'is the proposed project feasible?'

The architect should have looked at the client's 'want' and decided whether it was feasible or not at the introduction stage. If the architect considers the project is feasible then a contract or letter of instruction should be confirmed. The letter of instruction can be issued by the client to the architect or as a confirmation from the architect to the client. It is very important that the terms of the contract are clear. There should not be any 'offers' only agreements. For example 'we can act as the on-site agent' is an offer, whereas 'we agree to be the on-site agent' is a confirmation of the agreement. Alternatively, the RIBA and other professional bodies offer preprinted contractual documents for completion by the architect and the client. The contract clarifies to both parties the fees and consultants that will be employed to take the proposed project to the pre-construction stage. If, for example, the client does not want to proceed with the project after establishing an approximate costing, then the fees for the architect and the consultants will have to be settled. Any other costs such as searches or guidance from the Local Authority or other government departments would also be paid by the client.

This first contract is especially important. A recent legal case *Picardi* v. *Cuniberti* (Dec 2002) ended up in court after the client refused to pay the architect even though a CE/99 (conditions of engagement) and SFA/99 (standard form of appointment) had been issued, and a copy of the 'model adjudication procedure' as published by the Construction Industry Council. It appears that the architect had not taken the client through the various parts of the contract. The architect had written in a covering letter; 'Appointment. Please find enclosed the letter of appointment for your examination and signature. Should you have any query I will be happy to discuss it further.' The client had not formally accepted the contract therefore no agreement, no contract.

Feasibility is further broken down into two stages:

1 appraisal

and

2 strategic briefing.

The appraisal is carried out by the architect who would consider a range of issues before the investment of time and money continues. For example:

- Is the proposed project legal? The architect should know whether what the client wants to do is possible or not. Buying a plot of land in a mainly residential area and wanting to build a six storey office development is unlikely. If the area has other office developments it may be possible, therefore further advice will be required. There may be covenants or rights of way attached to the proposed land. Is the land planned for future development or could it be affected by future long-term developments?

- Is the client the legal owner or representative of the owner of the land or proposed developer? The client does not have to own the land at this stage. The intention may be to develop land which is for sale or the developer may want to know whether a project is viable before offering to buy land (see Chapter 8).
- An approximate costing will be needed to give the client an idea of the likely overall cost, although it is a guide only and is likely to be within about 25% accurate.
- The procurement method should be decided upon as it will have a major bearing on the procedure. The word 'procurement' has become associated with 'buying', but the Oxford Dictionary defines it as 'taking care of' or 'managing'. In the case of the appraisal, procurement will be the method by which the work will be carried out. For example:
  - Will the building contract go out to tender, and if so, what type of tender?
  - What contract will be used? A Standard Building Contract (SBC) where the main contractor will be responsible for procurement? Or a Client Management Agreement (CMA) with Client Management Trade Contracts (CM/TC) where the principal contractor (PC) will be responsible for over seeing the project? The PC will also be responsible for the procurement of shared plant, materials and labour, whereas the trade contractors will be responsible for their procurement. Will the contract be negotiated or billed? And so on. There are several alternatives (see Chapters 1 and 5).

Strategic briefing

Another word, 'strategy', perhaps clarifies the stage more easily. The architect and client will agree on the strategy for the project. Who will need to be involved? Which consultants should be used? What are the time constraints and financial constraints? Who will be responsible and for what? How should the project proceed and what is expected of the client? Does the client want the lead consultant to oversee the whole project and sign a certificate stating the work is of the agreed standard or just provide contract documents and obtain the various legal permissions? This would normally be stated in the contract.

4.3 Pre-construction period

Stage 2 covers everything up to the contractor being awarded the contract to build. By now the client would be aware of the likely costs and the team who would be carrying out the work of designing and costing. The architect would know whether he or she is the lead consultant or a team member as it will have a significant bearing on how the work will proceed. Like a ship needs a 'master' a project needs a lead consultant. The ship's master may not know the boiler pressures or how many eggs are needed in the galley as there is no need to become involved in the work; he or she would know that there is someone who is trusted to take care of the issue (procurement). Likewise, the lead consultant is likely to be the architect who will have a PQS to take care of the financial aspects and a structural engineer to take care of structural detail.

Outline proposals

The architect will develop the strategic brief into a project brief. The latter becomes a formal plan of approach prepared by the lead consultant in agreement with the client; concept design. The project brief will include:

- The appointment of the consultants will be confirmed and contracts written up with the lead consultant.
- A timetable for the whole project from design through to hand over of the finished project will be produced.
- The CDM co-ordinator will be appointed either by the client or the lead consultant.

Detailed proposals

The client will at this stage have agreed with the overall design and should authorise the lead consultant to proceed with the work.

Final proposals

The architect's design will become more detailed as to overall shape, colour, texture and location. Although many hours of work will have been put into the project it could still be scuppered by the Local Authority Planning Department. The client may have a title deed for the property and land, and the proposed work may enable lawful enjoyment of such land, but the Local Authority have an obligation under the Town and Country Planning Act 1990 to ensure that the proposed project will meet the overall plan for the area (see Chapter 7). If there is any doubt whether the project can proceed, an outline planning application should be made. If successful, the architect will continue by detailing the project. The detailing is commonly carried out by others in the practice – formerly known as architectural technicians, now referred to as architectural technologists. Their professional body is the BIAT (British Institute of Architectural Technologists). The other consultants will also work on the new detailed drawings. Today with CAD (Computer Aided Draughting) such as AutoCAD, all of the consultants can work on the same set of drawings as layers. It is quite common for the various consultants to be based many miles apart, even on the other side of the world, as all of the drawings can be updated electronically via satellite links.

When all of the details have been confirmed and accepted by the lead consultant and the client. the project would be submitted for a Full Planning Application with the Local Authority. It is common practice to submit the plans to the Local Authority Building Control Department, often referred to as the Technical Department. Alternatively another Approved Inspector can check for conformity with the Building Regulations.

Depending upon the method of procurement, if the project is to go out to tender the PQS will start a complete take-off before assembling a bill of quantities.

Production information

Although scaled plans, sections and elevational drawings would have been prepared to show compliance with the Building Regulations, there are many details that the builder will require to price up the tender and eventually build the project. For example, drawings of the window sections, fixings, glazing type and method, details of the handles etc., all drawn to a larger scale such as 1:10 or 1:5. The drawings would detail the position and type of the fixings, and weathering details. The specifications would be written for the various materials and elements contained in the design. The information would be passed to the PQS for addition to the tender documents.

4.4 Tender documentation

So far we have based this chapter on a traditional contract with a tender before the second contract with the building contractor is awarded. Tender documents comprise:

- scaled drawings showing the work to be carried out
- written description of the work to be carried out
- an outline timetable for the project stating the required completion time and conditions of the contract.

This is where the design stage finished. The next stages would be Tender Action and Construction Period.

In 2007 the RIBA updated the Plan of Work into the RIBA Plan of Work 2007. Although the stage headings have changed into five main groups:

- preparation
- design
- pre-construction
- construction
- use,

the content is very similar. The RIBA have in many ways simplified what is required at each stage (see Figure 4.2) compares the RIBA Plan of Works 1998 to the new 2007 version.

RIBA Plan of Works 1998	RIBA Plan of Works 2007
Feasibility	**Preparation**
A – Appraisal	A – Appraisal
B – Strategic Briefing	B – Design Brief
Pre-construction Period	**Design**
C – Outline Proposal	C – Concept
D – Detailed Proposal	D – Design Development
E – Final Proposals	E – Technical Design
F – Production Proposals	**Pre-construction**
G – Tender Documentation	F – Production Proposals
H – Tender Action	G – Tender Documentation
Construction Period	H – Tender Action
J – Mobilisation	**Construction**
K – Construction to Practical Completion	J – Mobilisation
L – After Practical Completion	K – Construction to Practical Completion
	Use
	L – Post Practical Completion L1 – Contract Administration during Construction L2 – Initial Occupation Services L3 – Review of Project Performance

Figure 4.2 RIBA Plan of Works comparisons.

The new approach to procurement has been further clarified using a flow chart colour coded to the main chart indicating the stage that the procurement issue is suggested. This is especially useful for students who are studying to become qualified RIBA architects.

For more information about the Plan of Works 2007 and the methods of procurement log on to the RIBA website at www.architecture.com/Files/RIBAProfessionalServices/ClientServices/RIBA%20Outline%20Plan%20of%20Work%202007.pdf.

In 1997 Public Private Partnerships (PPP) and Private Finance Initiatives (PFI) contracts came into being. A Government initiative to attempt to revive Britain's public services, the then Prime Minister Tony Blair was keen to encourage the private sector to join forces with the public sector in joint ventures. The theory was that the private sector would finance projects that the public sector would lease. Buildings such as schools, hospitals, and prisons would be designed, financed and built by the private sector and the public sector would then lease them on long-term contracts. (Doesn't it sound like a 'mortgage' though?) Critics of the scheme at the time said it would actually cost the taxpayers more in the long run and it would be better to continue with the public sector financing works in the traditional way.

Eleven years later, nearing the end of 2008, it appears they were correct. According to the BBC News online (12 February 2003), the Labour Research Department conducted a survey for the GMB union on the rent for PFI projects in the health service. They claimed it would top £13bn, equating to about £5 a year for every tax payer in the country – that was just the rent.

As a class activity, download the government report *Public Private Partnerships – The Government's Approach* published in 2000, and discuss the implications of the document with the benefit of hindsight: www.hm-treasury.gov.uk/d/80.pdf.

Were the critics right? Was the Government right? As a group, discuss the merits and demerits of whether it is better to borrow money and finance major projects or lease buildings on long-term contracts. The value of a pound today will be worth a matter of pence in 25 years' time. World influences affect banks and finance institutions, is it better for the Government to control major public building projects or borrow money to pay high interest rates and lease agreements over the longer term?

The advocates of PPP and PFI claim that the public sector cannot produce contracts or run projects as efficiently as the private sector. Why is this?

Advocates of PPP and PFI claim that without the schemes the hospitals, schools and prisons would not have been built. They also claim that performance-related penalties that appear in PFI contracts ensure the work is completed on time and on budget, something that public sector contracts could not achieve. Politics are relatively short-term by nature. Looking back at the historic traits, most policies are reactive as opposed to proactive. If a disaster happens, politicians want to appear to be responding to public opinion, especially as they could be voted out at the next election. For example, on 13 March 1996 one person fired a gun in a school and killed 16 children and their teacher. The news of the massacre in the Dunblane school gym outraged the press. The killer, a local man, Thomas Hamilton, who had a history of complaints investigated by the police regarding his behaviour around young boys. Nine years earlier, 14 random people had been shot dead by Michael Ryan in Hungerford in Berkshire. He reportedly turned the gun on himself and was found dead inside a school. The Government's reaction was to ban handguns in the UK from 16 October 1996. People who want to use handguns will continue to use them illegally with or without the ban, a fact that is borne out by the amount of gun crime still taking place in the UK. How many people died in road accidents in either of those two years? How many people died from lung-related diseases caused by smoking? As a politician, which issue would lose or gain votes?

A second topic for group discussion: are Building Regulations mainly reactive or proactive? The Government can change laws or amend laws when they want to, but when it relates to Building Regulations, can they be made to be enforced retrospectively? For example, all new buildings require sufficient thermal insulation to reduce the use of fuel/energy use. Would it be proactive to require all buildings old and new to have a minimum thermal resistance? At present grants are being made available for such work but there is no legal requirement to accept or improve the existing structures.

Chapter 5

Costings

As with all things the base factor is cost. Cost in time, cost to the environment, and cost in financial terms. At the time of writing the world is seeing the biggest turmoil in the financial sector since the 1920s.

Finance is a 'man made' costing; wealth. Before the invention of money, trade in articles, surplus food and possessions took place either willingly or by conquering. Today modern man equates everything down to the value of common factors such as gold. Money is just a token to be used instead of passing over quantities of metal. If the client wants a new building he or she will need to know the cost implications.

This chapter, albeit short, will give an insight into how the costings for a construction project are assembled. Before estimating became the norm the client would instruct the builder/designer to proceed without knowing how much the job will cost. The consequences included running out of money and the job being stopped, the workforce not being paid for their labour, or substantial changes in design and materials. A Scotsman first came up with the idea of estimating and tendering before the job was awarded to the builder. Prior to that, builders would often say how much cheaper they would have been after the job was complete. It was easy to say, but more difficult to prove especially before the event. Then came the job of the estimator – the chapter begins.

So far the architect has provided a graphical interpretation of what the client wants. A team of consultants have worked to the design and between them have produced a set of drawings and written specifications that should enable a contractor to build the project. However if the project is to have a bill of quantities prepared by the PQS a take off will be required.

In this chapter the fundamentals of taking off, preparing a bill of quantities and the techniques for estimating will be addressed.

Before starting the process of taking off, a plan for the whole building operation will be required, known as a 'method statement'. Divided into two main sections, by focusing on the main elements it should help prevent anything being left out of the bill.

- carcase
  - remove the top soil
  - excavate for foundations
  - foundations
  - external walls
  - internal walls
  - floors
  - roof
- finishings
  - windows
  - doors
  - fixtures
  - stairs/lifts etc.
  - plumbing
  - electrical
  - internal finishings
  - external works
  - drainage
  - pavements and footpaths
  - out buildings.

5.1 Standard Method of Measurement of Building Works

Although the method statement is determined by the 'taker off' (the person who reads the drawings, specifications and calculates the quantities) the list would be modified to suit the project in hand. Over the past four to five decades new building techniques and methods of construction have required changes to taking off. A standard method of measuring has been developed by the Royal Institution of Chartered Surveyors and the Building Employers Confederation to provide guidance to anyone taking off or estimating building work known as SMM7 (Standard Method of Measurement 7th edition). The book classifies the whole building process from Preliminaries/General Conditions through to Mechanical and electrical services measurement using a prefix letter and suffix number. For our example the SMM7 classifications

have been added to the method statement for ease of reference. (At the time of writing a new version will be available, SMM8, some time in 2009 or 2010.)
Carcase

Remove the top soil	D20 – Excavating and filling
Excavate for foundations	D20 – Excavating and filling
Foundations	E10 – In situ concrete
External walls	F10 – Brick/block walling
Internal walls	F10 – Brick/block walling
Floors	G20 – Carpentry/Timber framing/First fixing
Roof	G20 – Carpentry/Timber framing/First fixing

More detailed information relating to how to take off is provided in the sub-sections. For example, the topsoil removal would be [D20.2.1.1].

- D20 shows the main heading Excavating and filling
- The '.2' specifies excavating
- '.1' means the topsoil will be preserved
- The second '.1' is the average depth stated. Topsoil can vary in depth. Generally 150 mm is taken, but if the previous land use was ancient forest the topsoil could be considerably thicker. For our example the average 150 mm will be used.

The estimator will need to know what is to happen to the topsoil after it has been excavated. With planning law (Town and Country Planning Act 1990) topsoil and spoil may become part of the planning approval, requiring most if not all topsoil to remain on site and be incorporated in landscaping. Spoil from excavations may also have to remain on site as landfill sites are becoming more precious and there are fewer places to deposit spoil from building sites.

Example (you will need to use Figures 5.1, 5.3, 5.4 and 5.8)

Consider a simple L-shaped building with a strip foundation (see Figure 5.1). The plans show the external dimensions of the external walls. Although the building is L-shaped we would calculate as if it were a rectangle and deduct an area termed the want (the area that represents the missing corner is a want and if an area is within one of the sides it is termed a re-want). A sectional drawing would also be used (see Figure 5.8). Detail such as the wall construction, width of foundation, foundation type and materials should all be shown on the section. Note in this example the dpc is higher than normal. The Building Regulations require a *minimum* height for the dpc of 150 mm unless otherwise protected. In this example the designer has increased the height for two reasons:

1. To reduce the amount of spoil from the floor area (only the topsoil has been removed). The floor design incorporates rigid thermal insulation below the concrete oversite. It is good policy to increase the oversite thickness to 150 mm to reduce the possibility of the concrete cracking, especially from concentrated point loads from furniture. It is very expensive to rectify after the event.

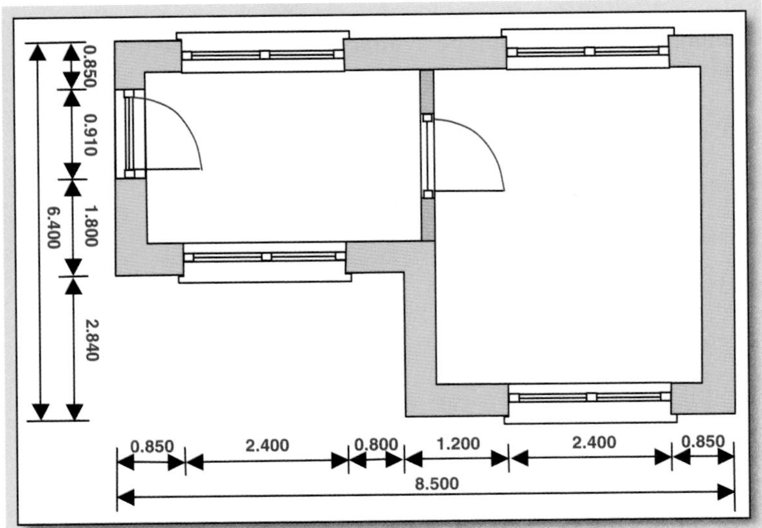

Figure 5.1 Plan layout [not to scale].

Continued

2 To help reduce the likelihood of water entry in flooded areas. The extra 150 mm in height obviously will not help if the flood water is 1 m deep, however, it may help in some circumstances.

With the sectional drawings a specification may be included on the drawing under 'Notes' or as an additional sheet. For example, special cement may have been specified if there are sulfates in the ground or ground water. The structural engineer may have specified steel reinforcement bars to be placed at the bottom of the slab if the building is on a hill or there is a possibility of ground movement. The type and size of steel including the number of bars etc. would all be specified. In our example the foundation is a simple trench fill in clay soil with no known sulfates therefore using Ordinary Portland Cement (OPC). The concrete mix has been specified at ST1. (A full explanation of the concrete design mixes can be found at Cooke (2007), p 182).

The previous classification stated the topsoil is to be 'preserved', therefore another set of information codes are required [D20.8.3.2].

- D20 – the main heading again
- '.8' is disposal
- '.3' is excavated material
- '.2' means it will remain on site.

The volume of excavated material will be given in cubic metres without any allowance for bulking. Also the distance for the temporary spoil heaps should be stated. The information will have a bearing upon the type of vehicles required to move the spoil and the time required.

If taking off in the traditional manner by hand, special sheets known as dim paper are used. A4 in size, the sheet is divided in two. Each side has 3 narrow columns and 1 wider column (see Figure 5.2). Working left to right the first

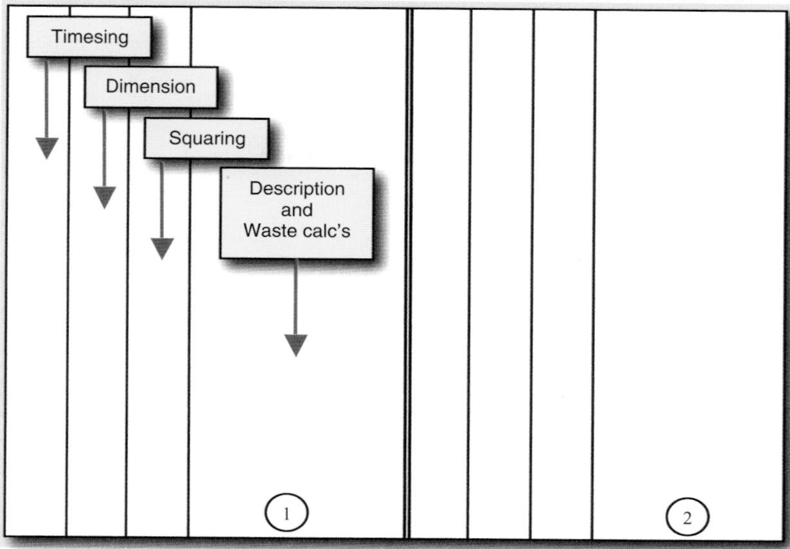

Figure 5.2 Dim paper layout.

column is the timesing column. Where more than one of an item is required it can be indicated in that column. The second column is the dimension or superficial column and the third is for squaring dimensions. Calculating the information should be carried out in the wide column. Hopefully it will become clearer as we go along.

The method of approach is to number each wide column at the bottom and encircle so it cannot be mistaken for a waste calc (see Figure 5.2). It is good policy to write a job number at the top right hand corner in case the dim sheets become mixed with other jobs when a file is used.

Using the method statement the first item is to remove the topsoil, so write a heading and description at the top of the waste column to show what the calculation will be for. For example: 'Projection Calculation' and underline it to show it is a heading. In most cases a short description of the task and additional detail is written below the heading. To link descriptions an ampersand sign should be used. It looks like a 3 written back to front with a near vertical line going through it. It is quick and easy to write plus it does not distract from the descriptions (see Figure 5.3). Abbreviations are used to reduce the amount of writing in the waste column. For example:

Abbreviation	Meaning	Explanation
a.b.d.	as before described	Where a description has been given previously it is easier and clearer to writer a.b.d. than to re-write the description.

n.e.	not exceeding	Commonly used in foundation descriptions enabling an approximate figure to be used with a maximum limit.
n.l.t	not less than	Commonly used to describe how
n.m.t	not more than	far an item or material should be moved.
av.	average	Also termed the 'mean'

Looking at the plan drawing, the external wall dimensions have been shown. However, the foundations project beyond the walls (see Figure 5.8), therefore a calculation to find the projection is required. The wall comprises a brick 102 mm wide + cavity 75 mm + block 100 mm = 277 mm overall. The foundation width is 450 mm, so subtracting the width of the wall from the width of the foundation will give both projections, then dividing by 2 will give 1 projection. The calculation is written close to the right side of the waste column to enable further detail to be written as required. To highlight a procedure such as adding several figures, the word 'add' with a wavy line beneath is written just above the figures to the left (see Figure 5.3). Any numbers written below the sub-heading will be added until a change of sub-heading therefore all of the main wall components will be added together and the result should be double underlined to show it is a final answer. A single underlined number indicates a sub-total. You can name the components with abbreviations for easy reference later.

To subtract numbers the term 'deduct' is used and abbreviated to ddt with a wavy line below. The abbreviation is written at the top of the calculation in line with the previous add sub-heading. All figures written below the sub-heading will be subtracted from each other. The sub-total is underlined singularly. To divide a number in taking off we multiply. For example to divide 100 by 2 we multiply 100 by $\frac{1}{2}$. To confuse matters more, we use a forward slash to show multiplication. In the example the sub-total is 0.173 (we always work in metres) with a forward slash to the left indicating that the number or fraction to the left will be multiplied. Note the way the fraction one half has been written, the line must be horizontal so that it cannot be mistaken for 1 times 2. The final result is underlined twice with the title to the left. P = 0.087. 'P' in this case stands for projection, which is 87 mm.

To calculate the area of topsoil to be excavated a new heading is written and underlined. Beneath the heading is the description of the task and the average thickness of the excavation. This is important as there is a set rate for specific thicknesses of excavation based on the type of plant required. An ampersand connects more detail indicating how far the spoil has to be moved. Again it is important indicating whether hauling plant is required or just a loading shovel or dozer could be used. In this example a want has been shown. A.b.d (as before described) the taker off will calculate the whole rectangle and then deduct for the want.

The calculation will be an area based on the length of the building plus the projection each end for the foundation (see Figure 5.3). Note the 'add' with a

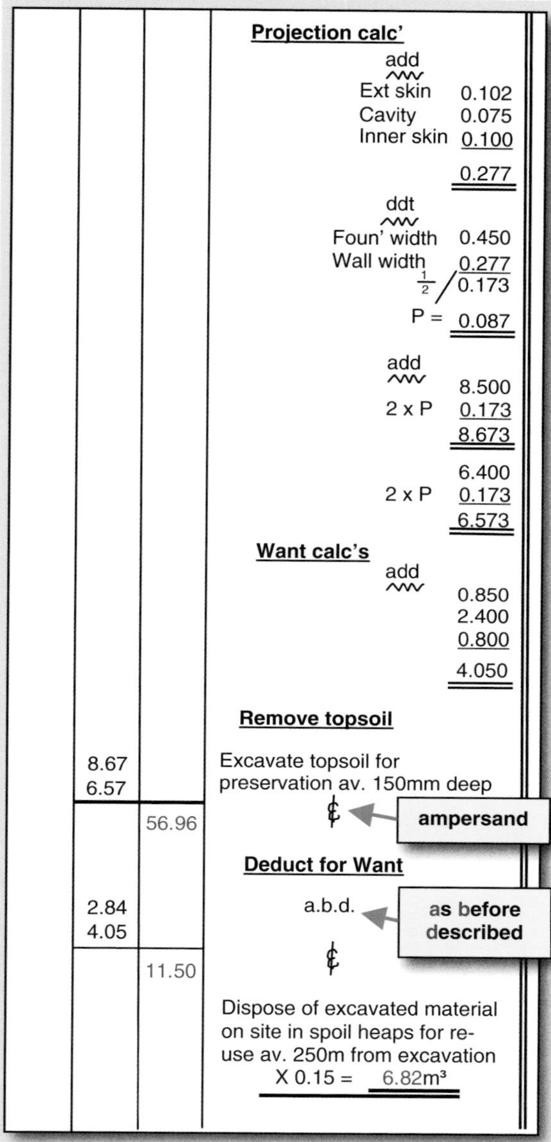

Figure 5.3 Topsoil calculation.

wavy line and the figures below have been added together with the answer double underlined for clarity. The same procedure for the width follows on all under the 'add' sub-heading. Now the calculations have been completed the totals (figures that were double underlined) are written in to two decimal places in the squaring column and a single underline across both columns in-line with the top of the description. A second underline goes across the third column as shown where the worker up will complete the calculation (shown in blue). The figure is an area and thus in metres square. The worker up will deduct the area for the want and multiply by the average depth 0.150 m and the total is written below the description, in the example 6.82 m^3.

5.2 Centre line calculations

The next stage is to calculate the foundation excavation. The technique is known as a centre line calculation. In our example the want in the corner can be ignored (see Figure 5.4). If the sections labelled 'A & B' are moved to the outside the total length of foundation is the same as if it were a rectangle. Note there are only four corners to make adjustments for. The centre line is calculated as the average or mean dimension. The measured length of the inside line shown in green is added it to the outside line shown in blue and the sum divided by two to give the average length, the same as the red centre line.

The blue line	$= 8.673\text{ m} + 6.573\text{ m} = 15.246\text{ m} \times 2$	$= 30.492$ m.
The green line	$= 7.773\text{ m} + 5.673\text{ m} = 13.446\text{ m} \times 2$	$= 26.892$ m.
The red line	$= (30.492\text{ m} + 26.892\text{ m})/2$	$= 28.692$ m.

Alternatively we could use the centre line technique. We have already calculated the external dimensions and double underlined them for the topsoil calculations. Using the total dimensions (double underlined numbers) adding

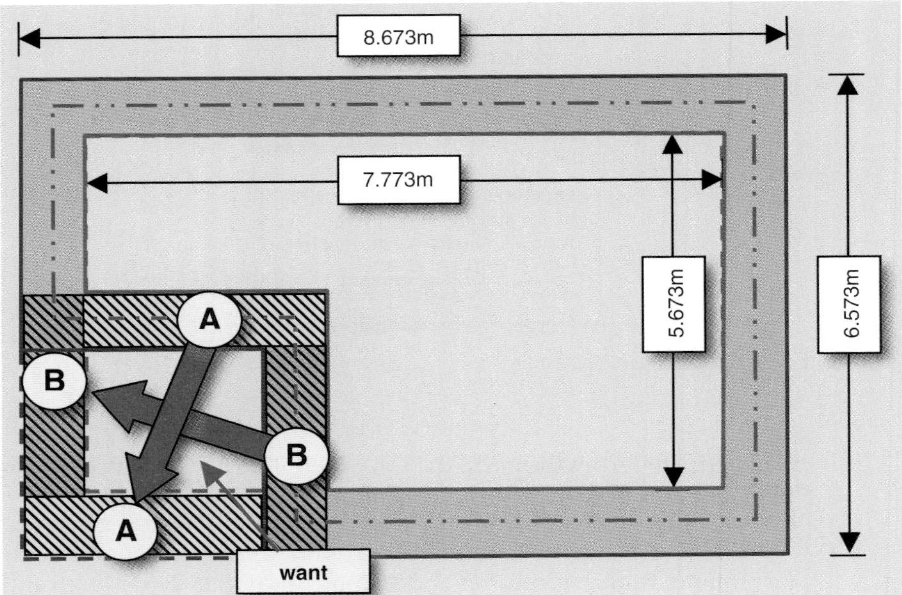

Figure 5.4 Deductions for a 'want'.

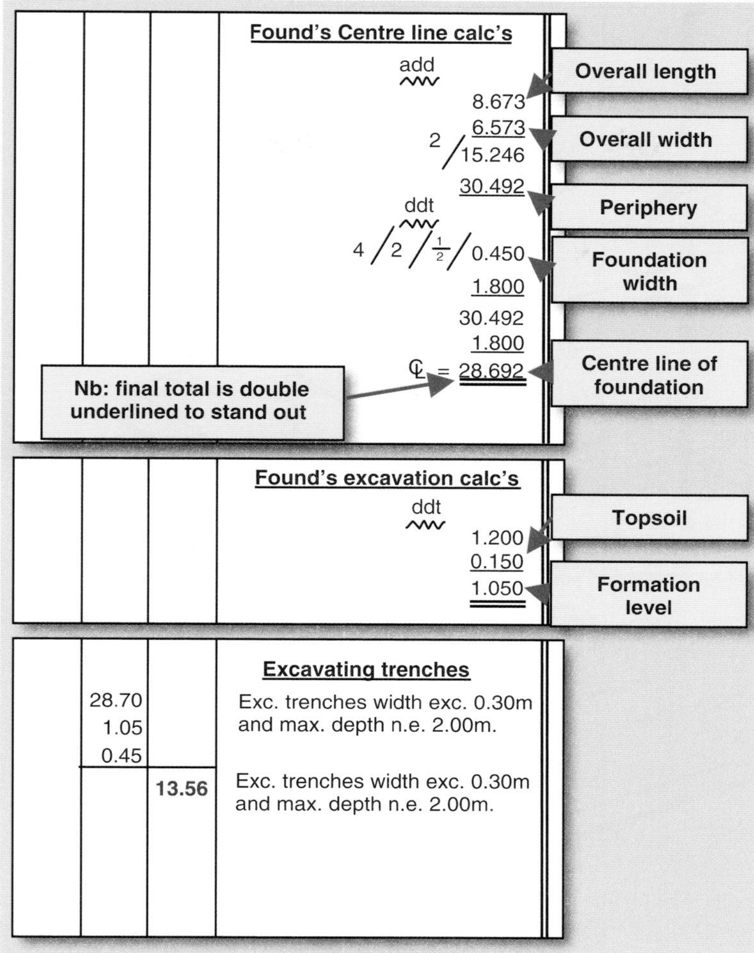

Figure 5.5 'Waste' calculations.

them together and multiplying by 2 equals the periphery of the foundation (see Figure 5.5).

To calculate the red centre line deduct half of the width of the foundation – see Figure 5.6 'A' and Figure 5.5 for the waste calculations. (To deduct half of the width multiply by $\frac{1}{2}$.) There are 2 sides to each corner therefore multiply by 2 (Figure 5.6 'B'), and in this example there are four corners therefore multiply by 4. Hence $4 \times 2 \times \frac{1}{2} \times$ width would be shown as $4/2/\frac{1}{2}$/width. The total is then deducted from the periphery and identified as the centreline by the letter 'C' with an 'L' going through it (see Figure 5.5).

The centre line for the foundation will also serve as the centre line for:

- foundation excavation [D20.2.6.3]
- earthwork support (if required) [D20.7.2.1]
- forming cavities in hollow walls [F30.1.1.1]

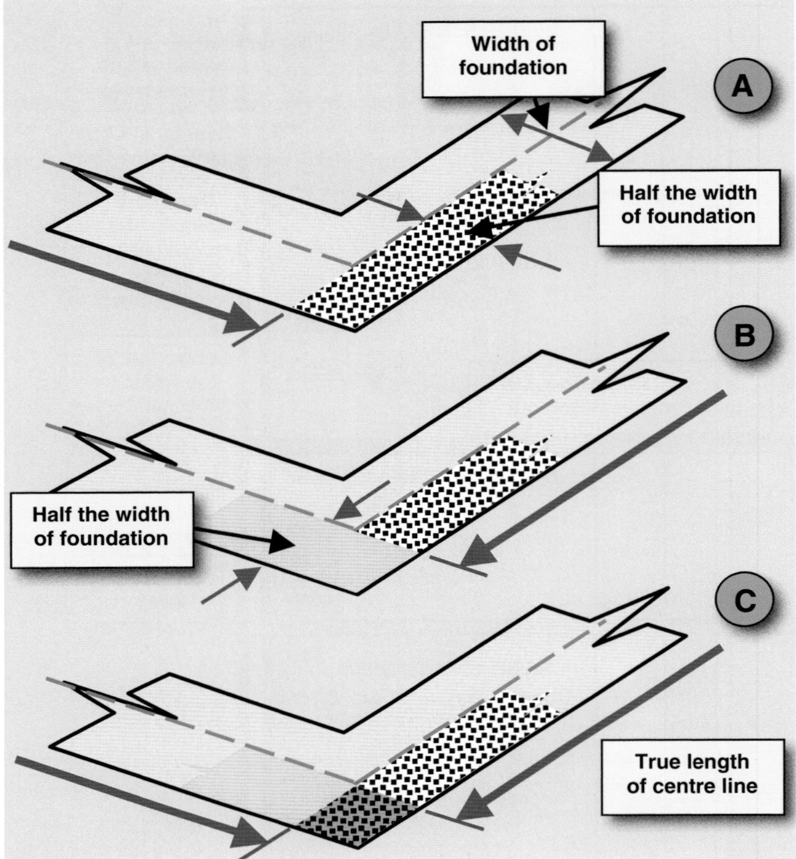

Figure 5.6 Centre line deductions.

- surface treatment [D20.13.2.3]
- lean mix fill [E10.8.1]
- cavity insulation [P10.3.1.2.M1.D1a].

Now the centre line has been calculated it can be thought of as the length of foundation required. The next dimension to be calculated is the depth based on the ground level (including topsoil) down to the formation level (formation levels are the point at which the natural substratum meets the built materials, in this case the bottom of the trench). However, the topsoil has been removed over the work area therefore it must be deducted from the excavation depth (see Figure 5.5). The calculations are complete, therefore a new underlined heading followed by a description providing lower and upper limits such as the width of the trench will exceed 0.30 m in width (0.450 m appears on the section) and will not be deeper than 2.00 m (1.20 m appears on the section). At this stage there is a small amount of topsoil to be replaced (termed re-instated) above the foundation to cover the projection. Although in this example it is a small amount, it needs to be taken for. We know the width of the projection

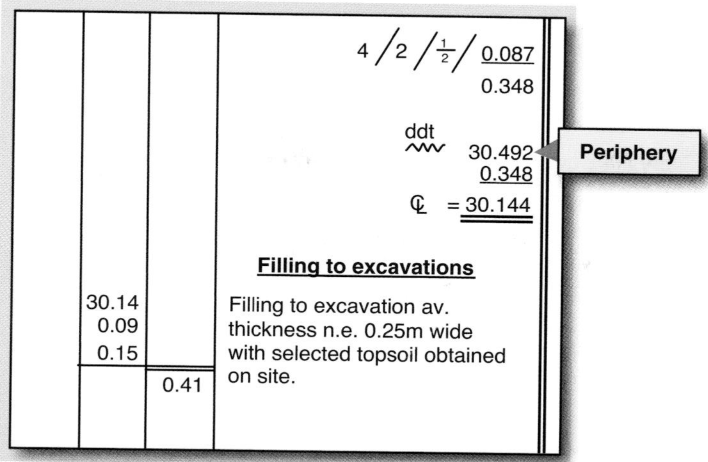

Figure 5.7 Centre line deductions 'waste' calculations.

is 0.087 m (see Figure 5.3), and the depth of the topsoil at 0.150 m, however another centre line calculation is required. Using previous information where possible, this is when the headings and double underlined totals are useful. For example, the foundation calcs show the periphery of the foundations. Using $4 \times 2 \times \frac{1}{2} \times$ width of the projection and deducting from the periphery, a new centre line can be established (see Figure 5.7).

Calculating the outer skin brickwork requires another centre line. Based on the periphery of the foundations, a set of deductions in a similar way to the previous re-instatement are required. The centre line of the outer leaf will be the width of the projection plus half the width of the outer leaf (in the example 0.087 m plus half of 0.102 m). The calculation should be shown in the waste column (see Figure 5.9). Another waste calculation will be required to determine the height of the brickwork from the top of the foundation to the dpc. When complete, include the description of the materials including brand name, product name, dimensions, and how they would be stated, including where they can be obtained or the manufacturer. In SMM7 the code would read F10.1.1.1:

- F indicates Masonry
- F10 more specifically Brick/block walling
- '.1' indicates the work will be to form a wall
- the second '.1' requires the thickness to be stated
- the third '.1' indicates the wall will be vertical and expressed in m^2.

Under the measurement rules M1 clarifies that unless stated otherwise the measurement is taken using a centre line. There are eight measurement rules in the section therefore identify the appropriate ones. The next column defines the work. For example, D2 states that facework is any work in bricks or blocks finished fair. That means the outer face of the wall will not receive any surface treatment such as paint, plaster or render. The 13 definition rules can be added

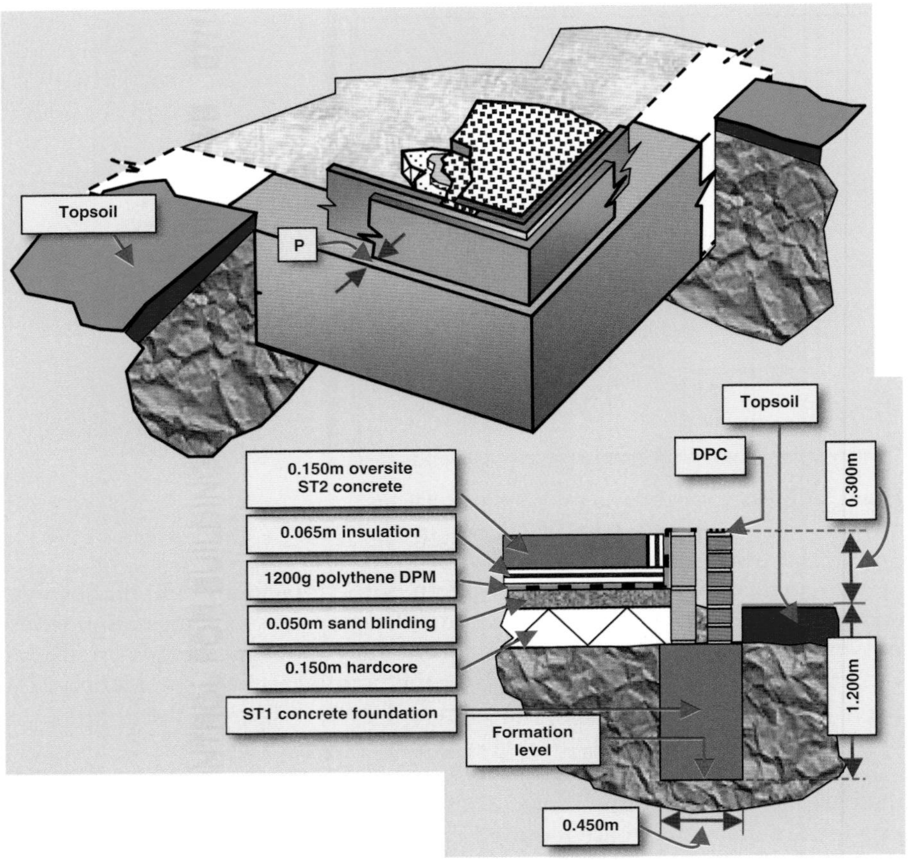

Figure 5.8 Substructure details.

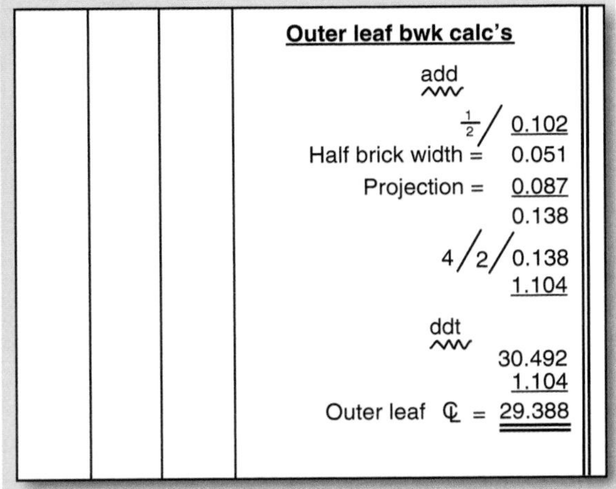

Outer leaf bwk calc's

add

$\frac{1}{2}$/ 0.102
Half brick width = 0.051
Projection = 0.087
0.138

4/2/0.138
1.104

ddt

30.492
1.104
Outer leaf ℄ = 29.388

Figure 5.9 Outer leaf 'waste' calculations.

where required to the list. Coverage rules define what should be included in the cost. For example, C2 states that the 'brickwork and blockwork bonded to another material is deemed to include extra material for bonding'. Where a wall has face brickwork at the end only and the main bulk of the wall will be concrete blockwork that will be rendered any extra material to form the bond will be deemed as included. i.e. cannot be added on.

The final column is supplementary information. S1 to S7 require extra information to be added such as S1 – the 'kind, quality and size of bricks and blocks'.

5.3 Walls in facings, half brick thickness

Taking off materials can be lineal either in metres (m) or metres run (mr), areas (m^2), volumes (m^3). Items such as damp proof membranes will be measured as the area they cover plus the area of upstand and the thickness of the wall. For example the plan area for the dpm will be the length of the external walls less the thickness of the outer leaf of brickwork plus the cavity and the inner leaf × 2:

$$\text{Length: } 8.500 \text{ m less } ((0.102 \text{ m } + 0.075 \text{ m } + 0.100) \times 2) = 7.946 \text{ m}$$
$$\text{Width: } 6.400 \text{ m less } ((0.102 \text{ m } + 0.075 \text{ m } + 0.100) \times 2) = 5.846 \text{ m}$$

written in the timesing column to two decimal places, thus becoming 7.95 m × 5.85 m, worked up to 46.51 m^2.

The upstand needs to be added. The inside perimeter will be the same with or without the want, therefore by adding the length to the width and multiplying by two the internal periphery can be calculated.

$$(7.946 \text{ m } + 5.846 \text{ m}) \times 2 = 27.584 \text{ mr}$$

The height of the upstand is needed and should be noted on the sectional drawing, in this case 215 mm (0.215 m). Therefore the upstand area will be:

$$27.58 \text{ m} \times 0.22 \text{ m } = 6.07 \text{ m}^2$$

The total area for dpm would be:

$$46.51 \text{ m}^2 + 6.07 \text{ m}^2 = 52.58 \text{ m}^2$$

Note that the SMM7 classification requires the pitch to be stated, hence *horizontal* for the main area and a separate figure, *vertical* for the upstand between the innerleaf and the concrete oversite insulation (see Figure 5.8). Coverage rule C1 *Work is deemed to include:* (b) bending and extra material for laps. The dpm would be taken up the innerleaf upstand and bent over the inner leaf to finish at the cavity. A dpc would then be placed over the dpm to ensure a good barrier to moisture rising through capillary attraction through the inner leaf. However, from a costing/estimating point of view the SMM7 considers the extra material as a lap therefore deemed to be included. In real life when the dpm, especially 1200g, is being pulled over the innerleaf before the continuation of masonry it is very difficult to get the corners neatly covered so the

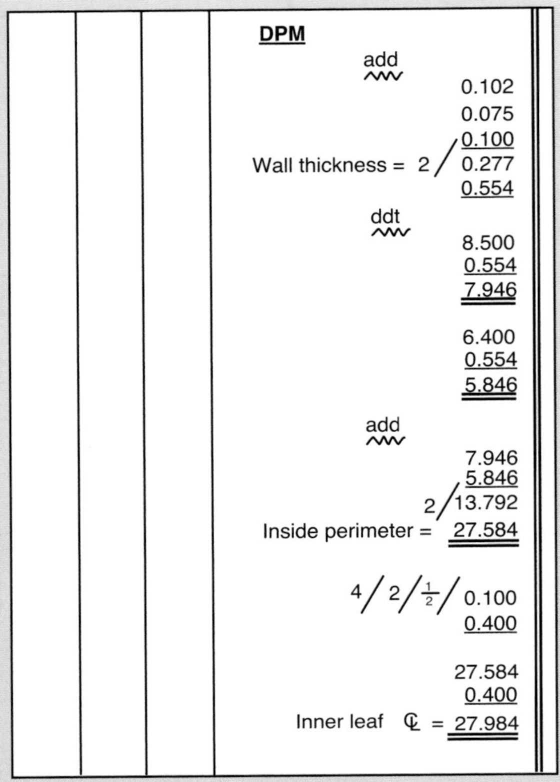

Figure 5.10 DPM 'waste' calculations.

corners are cut and excess dpm removed to enable the dpm and dpc to lay flat therefore the addition of a dpc is essential.

An adjustment for the want is made by a deduction. An ampersand with the letters a.b.d meaning as before described is followed by the want dimensions read directly from the drawing without any alteration. In the example:

Length: (0.850 m + 2.400 m + 0.800) = 4.050 m

Width: 2.840 m

The waste calculation has been previously calculated on a dim sheet (see Figure 5.3 and the drawing Figure 5.1). (When you are preparing the dim sheets it is useful to jot down the column number of the dim sheet for quick reference and checking. In the example the column numbers do not appear therefore numbers have been used.) The dims are written in the timesing column to two decimal places thus becoming 4.05 m × 2.84 m, worked up to 11.50 m^2.

All of the calculations would be written as waste calcs (see Figures 5.10 and 5.11).

Note that the description for the dpm has included a specific brand of polythene and thickness. As the sheet is very thin it is measured in microns. A

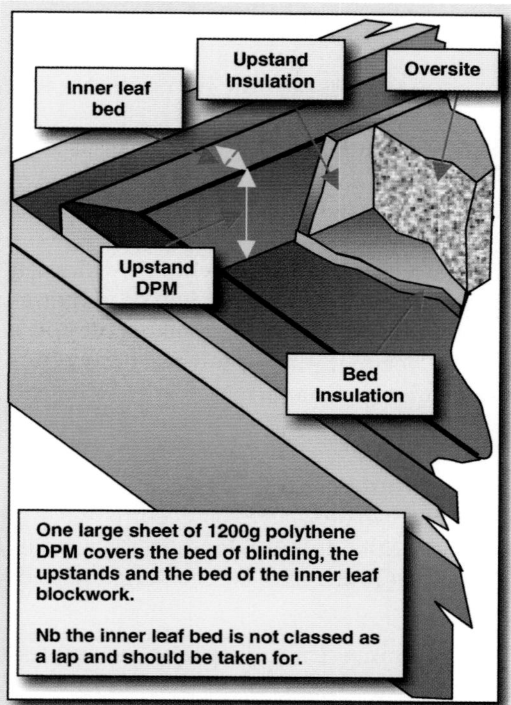
Inner leaf bed
Upstand Insulation
Oversite
Upstand DPM
Bed Insulation
One large sheet of 1200g polythene DPM covers the bed of blinding, the upstands and the bed of the inner leaf blockwork.
Nb the inner leaf bed is not classed as a lap and should be taken for.

Figure 5.11 Isometric view of the solid floor DPM.

micron is 1,000,000th of a metre, so 300 μm (μ is the ancient Greek letter mu, used to indicate microns) equals 0.0003 m thick or 0.3 mm. Plastics and other very thin materials are traded as per their gauge. 300 μm equals 1200 gauge, therefore to reduce the wording in the description 1200g can be used.

If there is any doubt about how to describe a product, a useful source is the manufacturer's technical pages and the Approved Document of the Building Regulations. In the case of damp proof membranes used on a ground floor slab they should be fitted in compliance with a Code of Practice, CP 102:1973 *Code of Practice for the protection of buildings against water from the ground.* The old method of stating it should comply with BS 743:1970 *Specification for materials for damp proof courses* relates more to the product than how it should be used, so the manufacturer and the Approved Document only refer to the CP 102:1973 plus amendments.

The inner leaf is calculated on a centre line multiplied by the height of the leaf from foundation to dpc level. It can be calculated by adding to the inside perimeter calculation used for the upstand and adding half the thickness of the inner leaf.

Thus 27.60 mr add half the thickness of the inner leaf times 2 times 4, which would be:

$$4 \times 2 \times \frac{1}{2} \times \text{ width of the inner leaf } = 4 \times 2 \times \frac{1}{2} \times 0.100$$
$$= 0.400 \text{ m}$$

Centre line of the inner leaf : 27.584 m + 0.400 m = 27.984 mr.

So far we have considered:

- How to calculate excavation trenches for the foundations using the centre line method.
- Using a single underline to denote a part total and double underline a full total.
- Double underlined figures are generally rounded up to 2 decimal places for use in the timesing column.
- Waste calcs should be headed up describing what the calcs relate to.
- The use of 3 decimal places and all dimensions will be in metres run (mr), metres squared (m^2), metres cubed (m^3). Some takers off will use a brick dimension of 102.5 mm (4 decimal places) but fractions of a millimetre on lineal dimensions are rather pointless.

We have also considered:

- How wants affect plan areas but not centre line calcs.
- Vertical areas such as walling can be written as a centre line multiplied by the height or, as in the case of the dpm upstand, a surface area based on the inside face of the innerleaf multiplied by the height from foundation to dpc level.

The principle for calculation is common throughout the take off for the walls, cavities, insulation, oversite and so on, right through to the wall plates.

- Plastered walls will use the same waste calc as the upstand lineal multiplied by the height of the wall (termed storey height).
- If the structure has more floors and the plan shape is constant to all floors, the waste calcs can be re-used each time. Remember to write a note to remind you and the checker which column the calcs appear on, otherwise it may take as long to find it again as to do the calculation.

The procedure for taking off window and door openings is to calculate the walls as if the openings were not there. Then deduct the area of each window in m^2 and state Deduction from walling 'a.b.d'. That saves having to write out the full description again. Additional materials will be needed such as vertical dpcs and or cavity closers, a dpc below the window, and additional wall ties as the centres are more close around openings. Depending upon the type of lintel a deduction for the area the lintel will take up in the wall may be required. However, with 'top hat' versions such as the IG L1/s, or the inverted tee lintels, no adjustment is needed as the blocks would sit on the steel tray (see page 231 of *Building in the 21st Century* for more detail on window and door openings including lintel types).

5.4 Bill of quantities

When the taker off had finished writing the bill it would be passed to the worker up (a person who would do the calculations such as multiplication, subtraction, addition etc. and enter the result in the squaring column). This is really history; when I first joined the estimating department calculators were just becoming available with basic functions and were very expensive. Slide rules were reasonably accurate and could be used to carry out many calculations including logarithms. However, all estimators had their own table book and adding machine that looked a bit like a typewriter – (they have the same model in the Science Museum in London). The book had pages of printed numbers with a column to the left that could be multiplied with the row of numbers at the top of the page, the result would be read from the numbers where the column and row met.

The dim sheets traditionally would have been copied, and then cut into the separate columns. Each description would have a heading and the SMM7 code written in squared brackets at the bottom of each description. To show where the description started and finished a squared bracket or slashed line would be drawn in to the left of the waste column (see Figure 5.12). When the columns had been cut the descriptions would then be cut just below the SMM code. Traditionally there would be a gap about the width of two

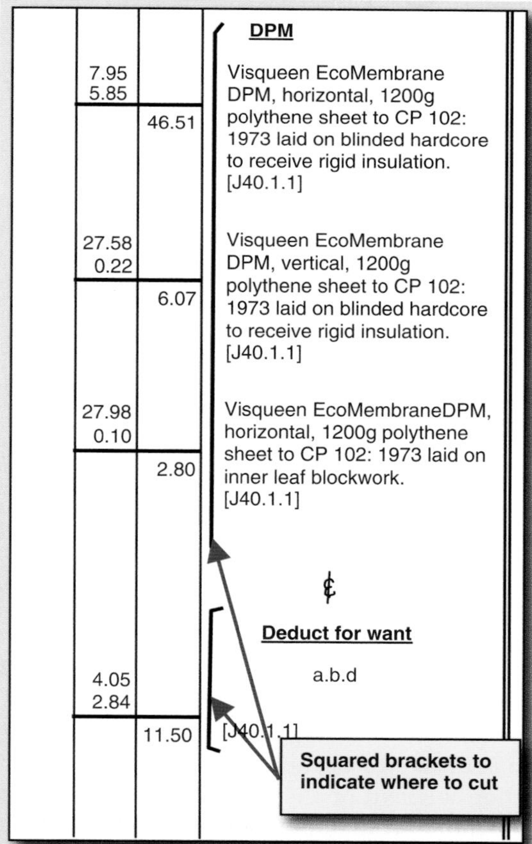

Figure 5.12 DPM 'waste' calculations.

fingers between the previous description and the beginning of the next one, thus plenty of space to cut each one out. The next stage was to sort all of the SMM codes into piles with the same type. For example, all descriptions with D20 would be placed in one pile and those with J40 in another. When the bill had been completely sorted (the method gave the name 'cut and shuffle') it would be given to the abstractor. He or she would then write all the descriptions in order under the trade headings.

As you may have gathered, it is very doubtful whether anyone still uses this method. However, in the 1970s it was the norm. The typist would type up the bill of quantities from the abstract sheets and either send it to the printers who would copy it up, bind it and professionally finish it, or if a very modern office with their own photocopier they would copy and punch a hole in the top left corner and fit a treasury tag. The method was very labour-intensive.

You may be wondering why has an historic process been included in a 21st century book? The BTEC syllabus requires its inclusion. A basic knowledge of how materials are quantified should help when using modern computer techniques. Taking off is still carried out, albeit with a 'magic pen' and pallet and very sophisticated software. The taker off selects the material either on the screen or using a pallet. He or she then traces the scaled drawing using a light sensitive pen and enters the third dimension using the keyboard. The software automatically selects the correct SMM reference and places the take off in the correct order. Pre-loaded or personally indexed descriptions can be selected, thus the whole bill can be produced and printed off by one person in a fraction of the time of the old method.

Note: When advising on costings the PQS should take into account whether less block or bricks are required, whether additional dpm above the lintel, additional insulation materials and not just the unit cost.

When the bill of quantities has been completed it usually forms part of the tender documents. When the client has a reasonably large project to complete, he or she may want to have several contractors submit their price for the work. The term 'tender' is used as the contractor tenders his quotation, known as the 'bid'.

5.5 Tendering

There are several versions of tendering that should enable the client to have the work carried out at a keen price and within the time parameters. They are:

- *Open tendering:* The project is advertised in trade and national press and any contractor can apply to tender. If the value of the project is above a specific limit, the European Directive requires it be open to include any contractor who is in the European Union. The advantage could be that a contractor who the client or the lead consultant had not considered or perhaps known about may put in a very keen tender. The disadvantage is that to issue tender documents is a very costly exercise for the client.

The BoQ can easily cost hundreds of pounds to print plus all of the drawings and specifications that form the tender documents. For that reason many clients require a significant deposit from the bidding contractors to offset the cost of issuing the tender documents and discourage the time wasters. Many reputable construction companies will not tender on open tenders as the odds of winning the contract are unknown. According to Brook (2005), examples of over 30 tenders have been known and an authority issued tender documents to 28 firms interested in tendering for a multi-storey car park using design and build.

- *Single-stage selective tendering:* This is by invitation only and not advertised. The client or lead consultant will formulate a shortlist of contractors that they either know or have had dealings with in the past. Commonly 6–8 suitable contractors will be approached and formally invited to tender. The project is given in outline and if the contractors are interested in tendering they will be given a set of tender documents. In practice contractors will not want to refuse to tender on the basis they may not be invited in future. Chief estimators usually have a good idea who the other contractors will be and how much work they have on their books. If, say, your company has a large job nearing completion and the proposed project will start about the same time, it would be common to put in a very keen tender price. In contrast, the contractor may know that the client has a reputation of frequently changing their mind and disputing all additional costs for extra work. In that case the policy may be to tender slightly high so that if the contract is awarded there will be enough extra money to offset the possible problems.
- *Two-stage tendering:* There are various versions of two stage tendering but due to the lack of space only one has been mentioned here. The first stage is based on an invitation to tender from the client or lead consultant. If accepted the contractor provides:
  - schedules of rates
  - the percentage for profit and overheads
  - detailed build up of preliminaries and
  - a plan of approach for the project,

  all based on the initial design scheme from the lead consultant. This is the first stage of the bid.

  Whichever contractor presents the most acceptable bid will be asked to price the BoQ when it is completed based on the rates previously presented in the bid. Depending upon the stage of design and the requirements of the client the contractor may be invited to become involved with the method of approach and to an extent the design using their expertise from the construction point of view. At this stage no contract has been entered into therefore if the project is aborted the client would only be required to pay for the work carried out thus far other than the initial bid. Some versions of this method of tendering will invite one contractor through to the second stage and have one or two standby contractors in case the first contractor wants to stop or the client and lead consultant cannot work with the contractor. Due to the high costs of tendering the number of contractors invited should not exceed 6. The Formal Invitation to Tender should state the actual number but not their names. For a more detailed account of the procedures and clauses it is worth reading: Code of Procedure for Two Stage Selective Tendering produced by the National Joint Consultative Committee for building (NJCC).

'Fee bidding' is a process where a contractor offers by invitation to carry out work based on a 'schedule of rates' or a price for the work. It cannot be a contract as such as there needs to be an offer and agreement to carry out the work. However, 'fee bidding' is a recognised method that enables the contractor and a potential client to establish whether it is worth proceeding further and pricing a Bill of Quantities with a contract. There is a significant time advantage as well. The client may select a contractor before the lead consultant has completed the design. Stage one of the 'two stage tender' will enable the cost consultant (CC)/PQS to look at the set of rates the fee bidding contractors have sent in. If the set of rates look close to those that the CC/PQS feel are suitable, then that contractor will be asked to complete a Bill of Quantities based on those rates when the design and 'take off' have been completed as the second stage. Note the word 'set'. The CC or PQS will look at the overall package of rates and not in isolation. It is not acceptable to either select some rates and not others or, worse still, communicate the lowest rates to one contractor in a Dutch auction. The rates are confidential between the client's representatives and the fee bidding contractor and must not be disclosed to a third party.

The advantages are:

- Speeding up the selection of the contractor thus the tendering process.
- A reduction in the amount of abortive work carried out by the non-successful contractors as they have not had to price a BoQ.
- A reduction in the amount of work and subsequent fees required by the CC or PQS in checking a selection of BoQ.
- The selected contractor from stage one may be invited to provide an input to the overall design from the contractor's prespective. For example, the techniques and plant that could be used during construction such as 'slip-forming' stair or lift shafts as opposed to 'cast and lift' the technique.

- *Serial tendering:* This is based on contractors tendering for similar types of contract. For example, Essex County Council architects department had a programme to build a number of new schools in the county. A system design was developed that incorporated specific materials and methods of construction such as large aircrete planks made by Siporex. Contractors tendering for one school would acquire specialist knowledge of handling equipment and time required to position, fix and finish the planks and panels. The contractor then could with more confidence quote for other schools using the same system.
- *Target cost:* In Design and Build Contracts the employer (client or lead consultant as agent of the client) will issue a target cost and a timescale

for the project in the tender documents that the bidding contractors should aim to match. If the lowest tender is greater than the target cost the contractor may be asked to consider alternative designs or methods to reduce costs. If the contractor feels that he/she would not like to reduce the design or costing, the second lowest tender may then be approached.

- *Measured term:* Specifically for clients who have programmes of regular maintenance and small/minor works. The client may be a large manufacturing company that has several old factory units. A programme of replacing all of the asbestos roofs with well insulated metal-clad roofing over a period of, say, 3 years. The client may have a maintenance department that would put the work out to a specialist firm using a JCT Measured Term Contract: 2006. The contractors would be advised of the scope of the work and the time parameters. They may be given a unit rate for the work by the client and the contractor would tender (place his bid) with a plus or minus percentage or at the unit rate if he feels he can still make a profit. The alternative method is for the contractors to provide a schedule of rates and the client can accept the most suitable.

Tendering for contracts is a very costly procedure for all concerned, both in time and money. Large complex tenders can cost companies tens or even hundreds of thousands of pounds to produce. The procedure is similar to a race – someone wins and the others get nothing – therefore it is important to select the work that is worth tendering for. Part of the tender process is to submit the completed tender in a specific envelope provided with the tender documents. The tender must be presented for a specific time, date and place, which is stated on the documents. Each envelope will be identical and must not have any identification marks such as a logo, post franked label or return address on it so that when the tenders are received they cannot be identified. Any identification marks will nullify the tender.

At the specific time all of the tenders will be opened and duly noted in order or total sum. Some clients allow representatives to be present at the opening, enabling assurance that everything is above board and no one's tender has been ignored. Usually, though, a minimum of two people open all of the tenders and note the final sums. The PQS, lead consultant and the client will then decide who will be awarded the contract. Local Authorities are obliged to accept the keenest tender, but the lowest price is not always the best option. Conversely, the most expensive does not necessarily mean they will be the best either. After careful consideration the successful contractor will be asked to submit their priced bills for further scrutiny by the PQS. The second and third keenest prices will be advised to be available. All the others tendering contractors will be told they were unsuccessful with their bid. All the bidding contractors will be informed of the winning figure and all of the other final sums without identifying the companies. Usually the chief estimators will know who is tendering for the job when they go to the merchants for materials or sub-contractors for their prices.

5.6 Estimating

Unit cost

There are several variations under this heading, ranging from, say, a complete bathroom down to a hand basin, taps, waste and fitting. When a PQS has, say, an hotel to price there may be 100 rooms, each with an en-suite. They may be handed (commonly en-suites will be back to back, thus reducing pipe runs and associated services). When one en-suite has been priced it can be totalled as a unit cost. Likewise, a bedroom can be costed as a carcase (the building part) plus finishings including furnishings. The total cost per room can be used, with or without an en-suite. The client and designer can then decide whether they require 100 rooms or whether the budget will only cover, say, 80 rooms. Another factor that can influence the number of rooms may be the Local Authority Planning Committee. Unit costs are very useful ways of costing and estimating.

Net cost

Look at the label on a jar of jam, or at the back of a container lorry. There is usually a weight shown. More accurately, at least two weights, gross and net (or nett). Gross is the whole weight, in the case of the jam it includes the jar, lid and label. The net weight is just the jam. In the case of the lorry the gross weight will be the maximum the container can contain including the weight of the container, and the net is purely the weight of whatever you want to put in it. Net cost is therefore the cost of an item or material only. It may be heavy, such as a parcel of bricks, so special lifting equipment is needed, but that will be an additional cost.

All-in hourly rates

This is not the same as an all-in rate. Where a worker is directly employed by the contractor a cost per hour should be calculated. It is not as simple as how much the person is paid.

Example

Fred is a site engineer paid £30,000 per annum. Fred had 25 days' holiday per year plus 8 days' bank holidays therefore 33 days per year.

33 days divided by 5	= 6 weeks plus 3 days
52 weeks in a year less 6.6 weeks' leave	= 45.4 weeks
45.4 weeks × 5	= 227 days p.a.

Fred works with an hour for lunch from:
8:00am – 5:30pm Monday – Thursday
8:00am – 5:00pm Friday = 42 hours per week.
45.4wks × 42hr/wk = 1906.8hr p.a.
£30,000/1906.8hrs = £15.73p/hr
Fred actually gets paid £15.73p/hr.

In addition there will be other benefits such as a company car, health insurance, life assurance, mobile phone, company pension, continual professional development (CPD), training and so on. Fred is likely to actually cost the company about twice his hourly wage. The company overheads would normally be apportioned to the contract sum rather than an individual employee. (It is something to consider when an employee leaves his or her job to have a smoke in the firm's time.)

All-in rate

In contrast with the above, the all-in rate includes labour, materials, plant, overheads and profit costs. For example, a small amount of concrete is required. It is not practical to have a part load of ready mix therefore the estimator will calculate how much 1 cubic metre will cost. The calculation will start with the basic materials: cement, sand, coarse aggregate. Plant is next; say a small cement mixer, three shovels, two wheelbarrows, and one rake hired for the job. Now labour; one on the mixer and two to barrow and place the concrete. One person will organise the work so will be paid more such as a ganger. (Various trades have different names for the person who controls the work, in this case it is a ganger.) It is most likely that the estimator will put all of the information on a spreadsheet so that when costs change he or she will not have to recalculate everything.

Unit rates

Unit rates comprise the cost of a material in a common unit of measurement including fixing. For example, skirting board is usually bought in lengths measured in linear metres or per metre run. The unit rate would be for 1 metre run fixed to the wall. Cutting to length and corners would normally be included in the unit cost. The build up of a unit rate is based on the SMM7. In this case P20.1.1 (skirtings) with coverage rule C1: the work is deemed to include ends, angles, mitres, intersections and the like except on hardwood items > 0.003 m^2 sectional area. Another example of unit rate is brickwork. This is measured in m^2 laid, i.e. the figure includes the mortar and wall ties where applicable.

5.7 Mensuration

On smaller projects such as one-off houses and extensions etc. or refurbishment work it may not be financially viable or practical to complete a bill of quantities.

Refurbishment work often entails removal of materials, walls or even rooms before the new works can be constructed, therefore estimators will complete calculations termed 'mensuration'.

Commonly area calculations are required. From area calculations volume calculations can be formulated. The important issues with estimating are:

1. Identify what the calculations are for – a simple heading underlined should suffice. (It is annoying and time-consuming trying to sort through sheets of calculations to find how the materials were calculated.)
2. Write the formula and carry out any transposition *before* entering the data – whoever needs to check the work will be able to see how you went about it. You may not be around at the time someone else needs to check your figures.
3. Keep the numbers neat and in the correct columns – it makes checking easier and less likely for mistakes.

The following section may help remind you which formulae should be used with a few worked examples.

5.8 Areas

The surface of any shape can be quantified as an area. The units of measurement will be in SI units (System International) mm^2 and m^2 for smaller areas and km^2 or hectares for larger areas such as land.

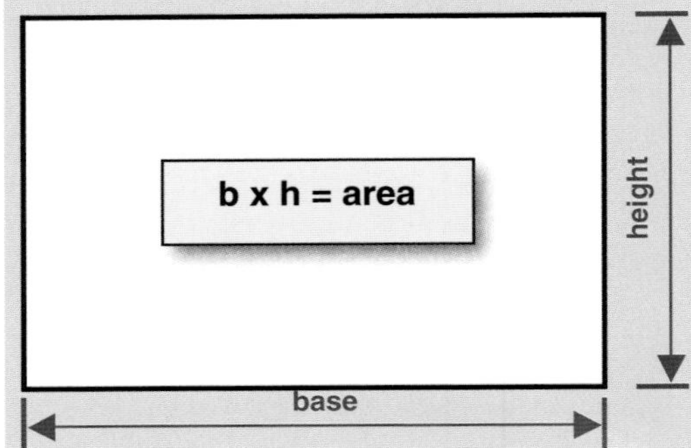

Figure 5.13 Area of a rectangle.

Areas of rectangles including squares

Formula: length or base × height (see Figure 5.13).

> *Example*
>
> A painter has to estimate how many cans of emulsion paint will be required to cover the wall area and ceiling area of a store room. The walls are to be magnolia and the ceiling white, and two coats will be required. The paint is available in 5 ltr cans giving an approximate 28 m^2 of coverage per can. The room dimensions are shown in Figure 5.14.

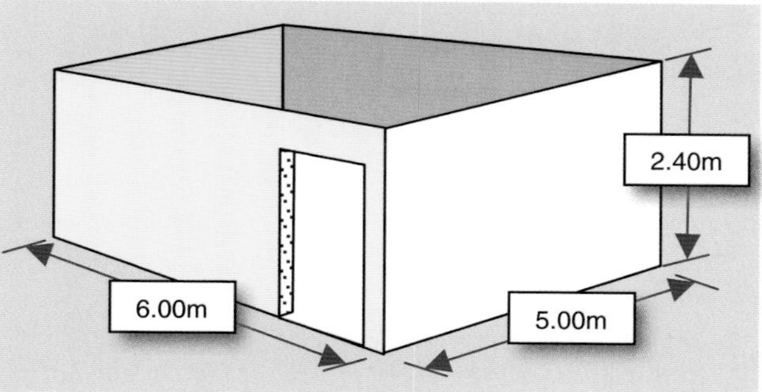

Figure 5.14 Volume of a room.

The doorway is 1.02 m wide and 2.0 m high and should be deducted from the wall area calculations.

Formula: length × height.

Stage 1: Wall area

Add all of the lengths together and multiply by the height.

$((2 \times 6.00\text{ m}) + (2 \times 5.00\text{ m})) \times 2.40 = 52.80\text{ m}^2$

$52.80\text{ m}^2 \times 2\text{ coats} = 105.6\text{ m}^2$

$105.6\text{ m}^2 \div 28\text{ m}^2$ (per can) = 3.77 cans = 4 cans of magnolia emulsion.

Stage 2: Ceiling area

$6.00\text{ m} \times 5.00\text{ m} = 30\text{ m}^2$

$30\text{ m}^2 \times 2\text{ coats} = 60\text{ m}^2$

$60\text{ m}^2 \div 28\text{ m}^2$ (per can) = 2.14 cans = 3 cans of white emulsion.

Notice both the long wall lengths are added first, then the two short lengths. Using parenthesis brackets '( )' around the numbers means that calculation is completed first. The outer parenthesis brackets indicate that when the two inner calculations are complete then they are calculated, in this case added. The order of carrying out calculations should be BIDMAS; **B**rackets, **I**ndices/**D**ivision, **M**ultiplication, **A**ddition and finally **S**ubtraction.

Areas of circles

Formula: πr^2 (see Figure 5.15).

π Ancient Greek letter pi = 3.142 or $\frac{22}{7}$

r^2 = radius × radius

Example

The area of an extractor fan has to be calculated. The radius is 75 mm.

Stage 1: Write down the formula

πr^2

Stage 2: Enter known data and complete the calculation

$3.142 \times 75\text{ mm} \times 75\text{ mm} = 17{,}674\text{ mm}^2$

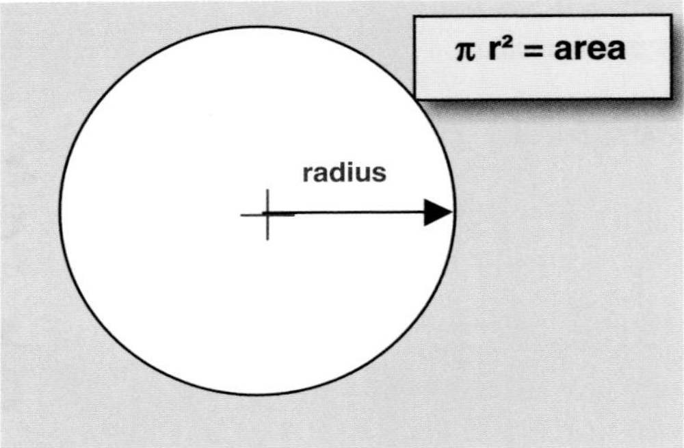

Figure 5.15 Area of a circle.

If you have a scientific calculator then use the 'π' button instead of 3.142. It is slightly more accurate 17672 mm^2

Surface areas of a sphere

Formula: $4\pi r^2$ (see Figure 5.16).

π Ancient Greek letter pi = 3.142 or $\frac{22}{7}$
r^2 = radius × radius

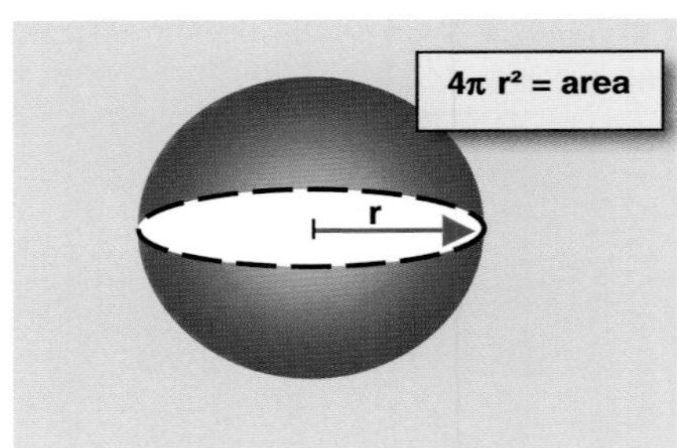

Figure 5.16 Area of a sphere.

> *Example*
>
> Stage 1: Write down the formula
> $4\pi r^2$
> Stage 2: Enter known data and complete the calculation
> $4 \times 3.142 \times 75\ mm \times 75\ mm = 70{,}695\ mm^2$

Using a scientific calculator = 70,686 mm^2

Areas of triangles

Formula: see Figures 5.17 & 5.18

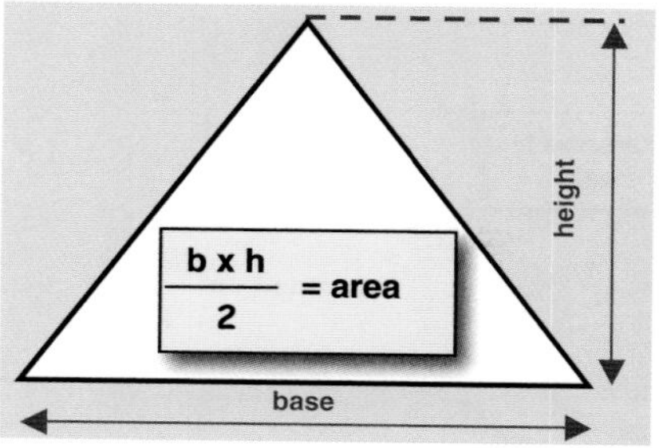

Figure 5.17 Area of a triangle.

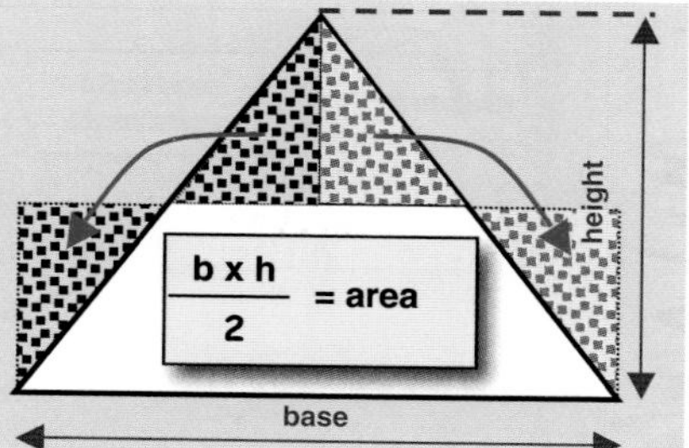

Figure 5.18 How the formula works.

Example

The bricklayer needs to know how many square metres of brickwork there will be in a new house (see Figure 5.19). He/she can calculate the wall areas and deduct the window and door areas as rectangles however the gable ends need to be calculated. Scaling the height dimension off the drawing it is 2.30 m high.

Stage 1: Write down the formula

$\frac{1}{2}$ base × height

Stage 2: Enter known data and complete the calculation

8.00 m × 2.30 m ÷ 2 = 9.20 m^2

There are two gable ends to the roof therefore 9.20 m^2 × 2 = 18.4 m^2.

If the pitch (angle of the roof from the horizontal) is known an alternative method of calculating the height of the triangle would be to use trigonometry.

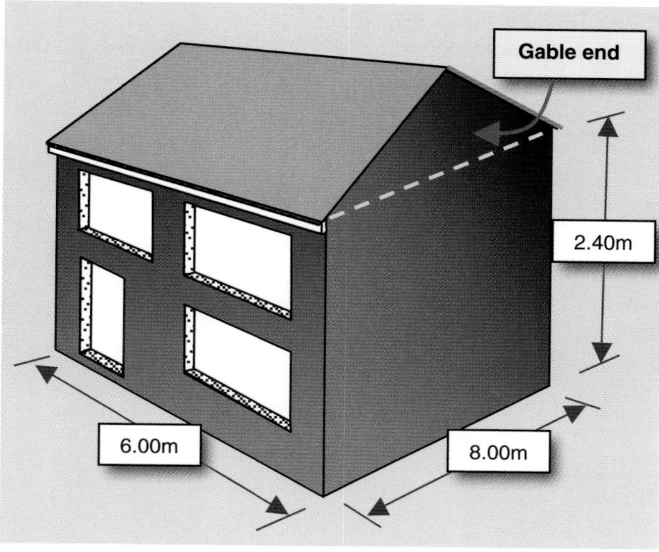

Figure 5.19 Area of face brickwork.

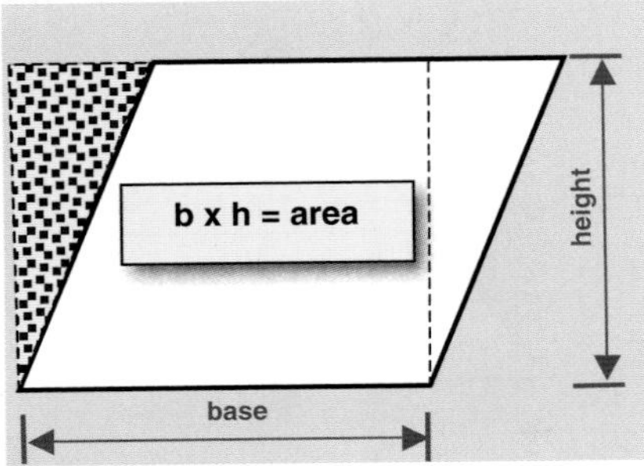

Figure 5.20 Area of a parallelogram without an angle.

Areas of parallelograms

Formula: length or base × height (see Figure 5.20).

If the base length and height are known:

Stage 1: Write the formula

Stage 2: Enter data and complete the calculation.

If the height is unknown but the two lengths and internal angle have been shown then trigonometry could be used (see Figure 5.21).

Trigonometry

There are three formulae that can be used with *right angled triangles*. A simple way to remember them is SOH TOA CAH, pronounced as 'sow to a car'. There are other ways of remembering the formulae, use whatever works for you.

Stage 1: Identify the sides (see Figure 5.22).

Look at where the angle has been shown as the sides always relate to the angle.

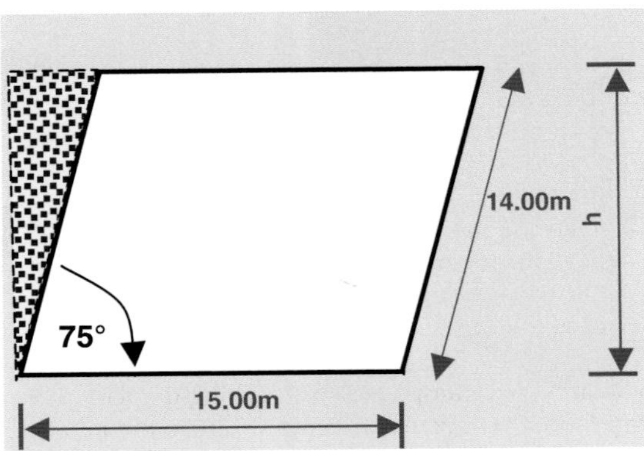

Figure 5.21 Area of a parallelogram with an angle.

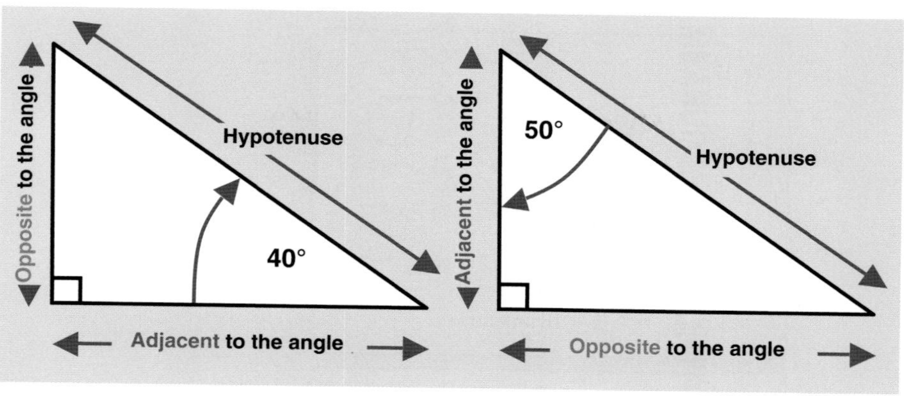

Figure 5.22 Trigonometry.

A small square in the corner indicates that angle is the 'right angle', it must be 90°.

The internal angle shown on the left triangle is 40 degrees. The side *opposite that angle* is known as the 'opposite' (opp). The longest side is *always* the hypotenuse.

That leaves the side next to, or *adjacent to the angle.*

Now compare the triangle on the right. The internal angle is 50°. Triangles must have internal angles totalling 180° (90° + 40° + 50° = 180°). The side opposite the angle is now at the bottom of the triangle, likewise the adjacent has changed.

Stage 2: Identify what data is known and what data is required and select the appropriate formula. (As long as the triangle has a right angle in it one of the three formulae will be useful (see Figure 5.23).)

Stage 3: Transpose (change the formula around) so that the unknown data is on the left of the equals sign (=) and the known data is on the right of the equals sign. To check the transposition is correct use some easy numbers, and transpose them.

Example

Figure 5.21 shows a shaded right angled triangle. We know the internal angle must total 90°, therefore 90 − 75° = 15°. The length shown as the unknown 'h' is the same as the adjacent side of the triangle. Therefore:

Stage 1: Draw the shaded triangle, identify all of the sides, show the right angle with a square in the corner and show the internal angle (see Figure 5.24).

Stage 2: Identify the known and unknown data and select a suitable formula. In this case we know the internal angle and the length of the hypotenuse. We want to find the adjacent length. CAH has the angle; Cosine 15°, the hypotenuse; 14.00 m, and the unknown side is the adjacent.

Stage 3: Write the formula and then transpose *before* entering the data. Complete the calculation (see Figure 5.25). Using the calculated length of the adjacent, enter it in the formula for the area of a parallelogram and complete that calculation.

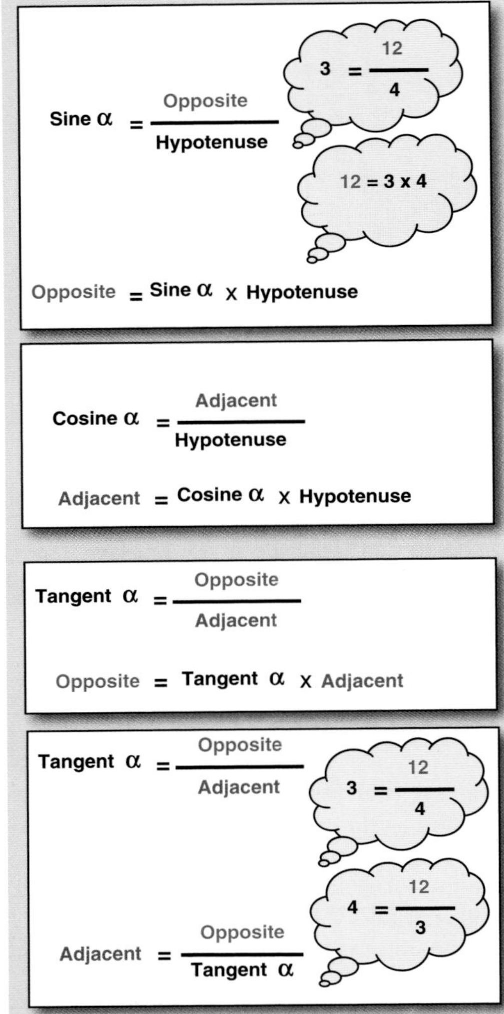

Figure 5.23 Trigonometrical formulae.

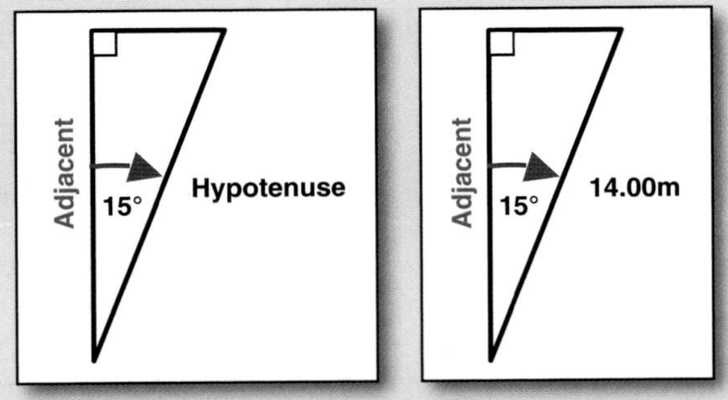

Figure 5.24 Identify the known information.

$$\text{Cosine } \alpha = \frac{\text{Adjacent}}{\text{Hypotenuse}}$$

Adjacent = Cosine α X Hypotenuse

Adjacent = Cosine 15° X 14.00m

Adjacent = 13.523m

Figure 5.25 Identify the correct formula.

The other formulae are used in a similar manner. Remember:

1 draw the triangle
2 identify the right angle
3 identify the known internal angle
4 identify the sides
5 identify and write the appropriate formula
6 transpose the formula; unknown on the left, known on the right of the equals
7 enter the known data
8 calculate and show the result.

Areas of polygons

If the shape is *regular* as shown in Figure 5.26 then count the number of sides and divide into 360°. If the flat side length is known, divide by 2 and that will be the opposite side to the internal angle. Calculate the adjacent length and multiply by the opposite length. Then multiply by the number of edges of the polygon.

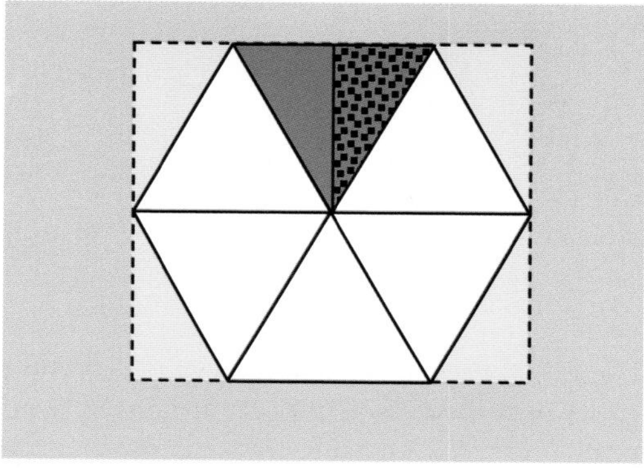

Figure 5.26 Area of a polygon.

$$\text{Tangent } \alpha = \frac{\text{Opposite}}{\text{Adjacent}}$$

$$\text{Adjacent} = \frac{\text{Opposite}}{\text{Tangent } \alpha}$$

Figure 5.27 Selecting the correct formula for the task.

Example

A hexagonal patio is to be covered in pea shingle. Calculate the area. The straight side edges are 3.00 m.

Stage 1: There are 6 sides therefore $360° \div 6 = 60°$. Divide by $2 = 30°$.

Stage 2: The opposite side to the internal angle is $3.00 \text{ m} \div 2 = 1.50 \text{ m}$

Stage 3: Draw the right angled triangle and write in the known data.

Stage 4: Select the appropriate formula based on the known and unknown data. In this case we know the internal angle, the opposite, but not the adjacent, so use the TOA formula (see Figure 5.27).

The calculation will be: $1.50 \text{ m} \div \tan 30° = 2.60 \text{ m}$.

Stage 5: The area of each right angled triangle multiplied by 2 will be the same as the area for a rectangle, length × height, therefore $2.60 \text{ m} \times 1.50 \text{ m} \times 6$ will be the area of the hexagon.

$2.60 \text{ m} \times 1.50 \text{ m} \times 6 = 23.4 \text{ m}^2$

Chapter 6

Stages of construction

Sometimes referred to as 'production', construction is the conversion of myriad tender documents into materials and the physical structure. Planning is essential and should be carried out at the pre-production stage, during production in the form of monitoring and adjusting where necessary, through to review for use on the next project.

This chapter guides through the strata of stages that will eventually culminate in the client taking possession. As with all stages of the construction programme, things sometimes go wrong: the importance of words on a contract both in meaning and in context have been shown with a selection of outcomes. Legal cases and the reasoning for the correct use of words and phrases will help to emphasise why time should be taken over the planning stages and why not to rush or gloss over contract procedures.

'Contra proferentem' is a legal term used when there is ambiguity in a statement, specification or written instruction. Where the contractor tries to place contra charges against the employer for running over the contract time based on claims of ambiguous instruction, a court will decide the meaning of the wording. Ambiguity (a phrase which could have more than one meaning) should not be confused with not understanding though. An ambiguous statement such as 'I have a pair of red and black socks' could mean I have a pair of red socks and a pair of black socks. It could also mean I have a pair of socks that are both red and black.

The chapter includes the various issues that must be arranged, ordered and programmed including methods of management and compliance with regulation.

There are six main stages of construction:

- design – (see Chapter 4)
- production
- maintenance
- alteration
- refurbishment
- demolition.

The definition of the word 'construct' is 'to make by fitting parts together, build or form' therefore considering the list above 'demolition' appears to be the odd one out. The design aspect has been addressed in Chapter 4. It will be further considered as part of Chapters 8, 9 and 10, where the role of the designer will be put in context of three different types of project.

6.1 Production

To 'produce' is the logical next step after design. The contract should be agreed in writing between the client and the contractor, or the client's agent (commonly the lead consultant) and the contractor. After the contract has been signed, including the agreed date for commencement, the contractor takes legal responsibility, known as 'taking possession', for the land, so the number one priority should be to make the site secure. No matter what state the site is in, if someone enters the land and hurts themselves or worse still is fatally injured, the contractor will be liable or will share liability with the land owner. It is no defence to state 'we have not started on site yet', or 'they should not have been on the land'. For example, an old shop may be on the land. It has stood empty for over a year, roof slates missing, rain cascading through the roof, every window broken by vandals. The shopfront had been temporarily boarded up but the vandals have pulled off the boarding at the back of the shop, allowing access. If, say, children inquisitively enter the shop and fall through rotten floorboards thus hurting themselves, the contractor would be held liable for negligence as the contractor had made no effort to make the site secure.

When taking possession of the site under a contract, the site becomes the domain of the contractor together with the responsibilities. The contractor is said to have 'possession under licence'. They should inspect the site and make it secure. Then if someone enters the site, as long as reasonable security has been achieved it will provide a defence against the claim.

In law the word 'reasonable' is frequently used. It is difficult to define what is reasonable. Brewer's Dictionary of Phrase & Fable, 16th Edition (1995) states that Sir Charles Bowen QC (1871–4) used the famous phrase in defence against a claim of negligence as 'the man on the Clapham omnibus' as someone who comes from an ordinary middle class part of south west London going about his everyday way of life. The definition is as 'what he would consider as reasonable'.

It is impossible to make a site 100% secure. If banks cannot achieve 100% security then 'reasonableness' is an effort short of 100%. For example, a 2.40 m high sheet hoarding around the curtilage (the outer edge or perimeter) of the site would be effective so long as there were no horizontal cappings or rails that could be used as footholds for anyone trying to scale the hoarding (see also Cooke (2007), p 357). In that example the hoarding comprised 18 mm thick plywood sheeting flush nailed onto softwood posts and rails so that there were no horizontal footholds to enable easy scaling. However, leaving the site door open and unmanned is not good practice.

Production can be considered in 3 parts:

1 pre-production planning
2 monitoring
3 post-contract reviewing.

6.2 Insurance

Taking out insurance for the contract is one of the first tasks of the pre-production phase. The insurance policy will cover the risk from a specific time and date and terminate at a specific time and date. The insurance underwriters need to know the duration of the risk and the amount of cover required. Consider the following scenario:

> A shopfitting company has entered into contract for the refurbishment of an existing shoe shop. The work will be carried out whilst the shop is open for business in two stages. The first stage will be the basement fit out followed by the ground floor fit out plus extension to the small existing store room. The contract starts from 0100 on 1 June 2008 and will last for 6 months. As work proceeds and nears completion the stock room has been completed and all the shelving, lighting and signage put in place. However, the smoke detectors and fire alarm have not yet been wired up to the main system. The manager of the shoe shop asks whether he can start getting the new stock in ready for when the shop opens in 2 weeks' time. The stock will have to be carried through the shop where the shopfitters are trying to finish off their work. The foreman fixer agrees to the request (The term 'fixer' is used in shopfitting for a craftsman who works on site. Basically, they fix any units, false walls, shelves and so on, hence the term 'fixer'. The foreman would be the chief fixer, who is responsible for the work done on site.)
>
> During the night a fire breaks out and destroys the whole shop including the new stock. Who is responsible for covering the claim for the total loss? Is it the contractor who worked in part of the shop? Is it the shop owner or manager? Whose insurance will be asked to pay out for the loss? As you can see, from a simple request from the shop manager and a helpful foreman a complex situation has arisen. The insurance liability should be sorted out at the formation of the contract stage. The new

works forming the contract under possession will be the responsibility of the shopfitter. Insuring the same property under two separate insurance policies is not allowed and will result in both insurance companies nullifying their policy. The correct procedure would be to have a joint insurance where the employer and contractor are both named on the policy covering existing work and new work. Alternatively, the contractor may be required to insure the new works and materials. This type of insurance cover is more suitable where areas of construction can be easily defined. It would not have been suitable for the shop refurbishment though.

On larger developments a term known as partial possession can be included. That means part of the project can be officially handed over to the client by agreement of the contractor. Documents are prepared specifying the extent of the work that is to be handed over and when the handover takes place. The client then will be liable to insure that part of the completed work and any additional contents therein. If the work has been a part of a JCT form of contract such as JCT 2005 Design and Build Contract (Revised 2007), then clause 2.3 relates to the partial possession by the employer. The clause is relatively brief in contrast to the former JCT contract Standard Form of Building Contract With Contractor's Design 1998 Edition. Even with a formal contract such as those prepared by the JCT, things can still go wrong. For example, in the court case *Impresa Castelli SPA* v. *Cola Holdings Ltd*, according to Harris of the solicitors Wright Hassall, Impresa Castelli agreed to build a large four-star hotel on behalf of Cola, the hotel proprietor. The contract had been for 19 months from possession by the contractor to practical completion where the work would be handed back to the employer/client. As there had been delays the contractor had asked for an extension of time of three months. The employer sought liquidated damages of double that in the contract from £5,000 per day to £10,000 per day. Further delays were experienced due to the air-conditioning systems which meant that a second variation agreement had to be made where the hotel would be fully operational by September instead of the contract date of February. The employer would be given access to parts of the hotel although not fully completed. In such a case the architect could, and should not issue a statement of certificate for hand over of the work as it was not as per the agreed contract, i.e. not completed.

The documents between the two parties included the word 'access' as opposed to 'partial possession', which meant that any claim for liquidated damages had been lawful under the relevant clauses. For an excellent précis of the case plus other cases of partial possession, visit the Wright Hassall website: www.wrighthassall.co.uk/resources/articles/art_construction07051.aspx

As you can see, even with a written contract things can still go wrong. The use of specific wording can make a lot of difference especially if the case goes to court.

For readers who are studying building law, the full judgment can be found at www.adjudication.co.uk/cases/impresa.htm.

As you can see, insurance is another very important issue that must be arranged prior to the commencement of the contract. Public liability insurance is required of all trading companies. It is also a good thing for all householders to have public liability insurance, as anyone who enters their property, such as the postman, paperboys or girls, could claim if they are injured by, say, a slate or tile slipping off the roof. An example of a real life roof problem and potential accident can be seen if you visit www.blackwellpublishing.com/cooke/teaching.asp.

The photograph used for Figure 2.3 in pack 2 (Surveying processes) shows roof slates on the roof of the reception office of a factory unit that is constantly in use. If an accident happened and it could be proven that the factory owners had knowledge of the poor state of repair and lack of maintenance they might be liable for negligence. If on the other hand the roof appeared to be in a good state of repair and there had been no previous slates slipping off the roof then the public liability insurance would settle the claim for compensation as the factory owner would have been reasonable. Although a simple example of reasonableness where large claims are being considered it is likely that a County Court would have to decide whether the factory owner had prior knowledge of the situation and had taken any steps to replace or carry out repairs prior to the accident occurring.

Insurance coverage is part of the contract. For example, in the JCT Design and Build Contract 2005 (Revision 2007):

- Clause 2.3 states the contractor must arrange insurance cover for the works from the date of possession to the date of completion.
- Clause 2.5.1 covers insurance where the employer requires early use of the works, i.e. before the practical completion date.
- Clause 2.22 relates to the insurance cover for any listed materials that have been included in an interim payment wherever they are until they become part of the work where the works insurance would cover the items. For example, a large stainless steel and glass shield has been made and will form the main crest above the entrance of a new office block. The work on site is running two weeks late and the contractor asks the shield supplier to deliver the item to a lock-up garage he is renting on the site next door. As the item has been supplied and delivered the contractor will be expected to pay for it therefore he has included the cost of the shield in as part of the interim payment. If for example an accident happens and the shield is damaged the insurance under this clause should cover it even though it was not on site.
- Clause 6.4.1.2: Contractor's insurance against damage to property or injury to persons: Whilst the contractor has possession he must indemnify the employer (meaning the contractor must insure the works and property in his possession) against any personal injury or death caused by the works. The only exception would be if the employer or employer's persons are proven to have been negligent.
- Clause 6.4.1.3: This sub-clause covers the eventuality of the contractor not taking out an indemnity. In such a case the employer can take out insurance and deduct the costs from the Contractor as a debt.
- Clause 6.5.1: Employer's liability insurance. The contractor may be required to include joint liability insurance cover with the employer. For

example, where the employer still operates from within part of the curtilage of the site, a joint liability insurance would be obtained. If any claims for damage caused by excavations, vibrations, heave to adjacent land and buildings are made as the work proceeds then the insurance will come into effect. Dewatering of a building site near St Paul's cathedral in London caused subsidence of the cathedral and works had to be stopped immediately. Now radial zones around the cathedral identify to what extent ground water may be removed.

- Clause 6.7 and schedule 3: Insurance of the works: The extent of the insurance cover is set out in schedule 3: Control of the works as one of three options: A – all risks insurance by the Contractor, B – all risks insurance by the Employer, or C – Employer's risk for the existing structures, and the Contractors for the new works. (Option C would be typical for an extension to an hotel, shop or other significant structure.) To ensure that the renewal date is not overlooked the Contractor must enter the appropriate date as part of Clause 6.7.
- Clause 6.11 refers to professional indemnity insurance. Specific risks would be noted that may include pollution or contamination claims. Unlike the insurance for the works professional indemnity insurance has to continue for many years after the completion of the contract. For example, if something like hydraulic fluid had been stored on site in metal containers inappropriately directly on the soil. The containers had rusted and leaked saturating the ground beneath them. Then the area was cleared and the contractor placed a concrete slab to form part of the car park. A year later the fluid migrated through the soil and contaminated the water course that lead into a local lake killing many of the fish. The land owner may have been able to trace the source of pollution back to the building site. Therefore the contractor must keep professional indemnity insurance to cover such events for several years.

In addition to the above there can also be three options for insurance cover. They cover such events as the shoe shop scenario where the employer wanted to place his stock in the stock room with the consent of the contractor. In contrast; Option 1 is suitable for new-build where the policy is an 'all risks cover', although 'all risks' will have a list of exclusions and may vary between insurance companies. For example the insurance will not cover a loss due to bad design, poor workmanship, or inadequate specification. Like all policies it is wise to actually read the policy documents *before* the insurance comes into effect to ensure the cover meets the requirements. It is of little use finding out later that a risk would not have been covered or had been excluded. Exclusions apply more to material items rather than accidents or injuries to third parties.

Even the annual renewal date for the insurance must be written in as part of the contract

So far we have made the site secure and arranged for insurance cover. Other pre-production issues include:

- arranging utilities
  - water

  - electricity
  - sewage
  - drainage
  - telecoms
  - cable communications
- applying for licences
  - temporary road closure
  - temporary footpath closure or re-direction
  - suspension of parking bays
  - temporary removal of street furniture
  - erection of hoarding
  - temporary advertising other than for the purpose of site recognition
- informing the Health and Safety Executive that the site will commence work (F10)
- informing the Building Control department that the site will commence work
- setting up accounts with suppliers.

6.3 Arranging utilities

Water

Utilities are the mains services that go to or across the site. For example, potable (drinking) water will be piped adjacent to the site in most cases in a large diameter pipe known as the mains supply. From the mains a *communication pipe* would be connected with a meter to measure the amount of water taken through the pipe. The Model Water Byelaws 1986 with amendment 1987 are a Government statutory document to regulate the supply of water. The bye-law was made to prevent waste and reduce contamination of the water supply. By doing so it is considered that misuse and undue consumption will also be addressed. For example, all external taps must have an anti-siphonage valve fitted to prevent external water being siphoned back into the mains if the water pressure falls. Prior to the new regulation a hose could be fitted directly onto a tap for, say, washing a car. If the hose was left in a bucket containing detergent it may become a siphon if the water pressure dropped, thus sucking the detergent and water from the bucket and back into the mains. The scenario could be similar where a hose pipe is left in any container of liquid. Taps within a building, such as those fitted to washing machines, dishwashers and water softeners, all require anti-siphonage valves to be fitted to prevent contamination of the water supply.

The contractor will apply to the Local Water Authority under section 45 and 55 of the Water Industry Act 1991. Examples of the documents can be downloaded from the Local Water Authority website. The information required includes:

- Fire protection details – whether a fire hydrant will be required, or hose reels etc.

- What usage the incoming water will be put to – fixed appliances such as washing machines, dishwashers, WCs, baths etc. Whether the internal system will be direct feed to the taps or indirect feed via storage tanks or cisterns. Also, whether the hot water will be via a pressurised system.
- The distance of runnage on site and whether there are any contaminates in the soil. They will also need to know any appropriate history of the land, both on the site and on adjacent property. For example if the land had previously been used as a coal gas works it is probable that many of the chemicals that remain in the soil could attack the MPDE blue pipe commonly used for water mains, in particular coal tar, cyclohexane extractable material used as adhesives, resins & rubbers, phenols (otherwise known as carbolic acid), polyaromatic hydrocarbons (tar), toluene extractable material (road surfacing & roofing material) and total petroleum hydrocarbons (more commonly known as petrol and motor oils). All of these substances can dissolve or migrate through the plastic over time, thus contaminating the water supply. For more detailed information go to: www.thameswater.co.uk and search for 'Drinking water standards explained'.
- What the expected demand for water will be. This would be calculated at the design stage by the services engineer (see Chapter 4).
- Although permission is required from the Local Water Authority to connect to their water main, the contractor may employ a subcontractor who is Water Industry approved to carry out the connection and appropriate testing. A declaration that the Local Authority has given Planning Permission for the works forms part of the application.

All of the above will need to be programmed in for connection. If additional or larger feeds are required, it may take months or even years to bring in the supply. The utility company would normally have been approached during the design stage, so lead in times should be reduced.

Electricity

Mains electricity may run over the site by way of cables on poles or pylons or, more commonly in built-up areas, underground. The contractor would apply to the DNO (District Network Operator), who under Government statutory regulation has a duty to supply connections to premises. A fee will be payable based on Licence Condition 4: Connection Charging Statement, which can be found on the DNO website. The application should have been made at the design stage when the services engineer calculated the electrical requirements for the project. The DNO will provide a written quotation, but the supplier of the electricity must be selected before the installation becomes active, preferably before accepting the quotation. The DNO will provide the connection from the mains supply adjacent to the site up to and including a metering device in the new work. In a similar way to the water supply connection the electricity supply connection can be sub-contracted out to an accredited electricity connection service provider. Any electrical switching gear or distribution boards will be part of the contractor's remit. If the project requires high voltage mains supply (110,000 V) the supply cable remains the property of the supply company up to the contractor's switching gear. For example on the Ropemaker site discussed in Chapter 10 the electricity supply company have their own electricity service room within the site. At the college where I work the electricity supply company have two fenced-off areas containing their service connections of 110,000 V (identified by the red outer casing to the cables) and transformers including cooling radiators.

Sewage

Most new buildings would be connected to the mains sewage network if the site is in a built-up area. However, not all new buildings have easy access. For example, in remote villages or farms, provision to store effluent and waste soil water (water that has been contaminated such as bathwater, washing machine and dishwater outlets etc.) on site has to be made. Although it would have been detailed during the design stage and as part of the Building Regulation Full Plans Approval, the contractor should make provision with the Local Water Authority. Where connection to the main sewer is required, an application to the Local Water Authority under section 106 of the Water Industry Act 1991 is required. As part of the application reference is made to whether Planning Approval has been granted for the project.

Drainage

Drainage, in contrast with sewerage, is the removal of surface water from a site. In many cases the surface water is piped into a combined system with the foul water connection previously mentioned. Alternatively, a separate system may be used. The Local Water Authority need to be advised of the method of surface water removal, even if it is to be run to a 'soak-away'. Where National Government are pressurising Local Authorities to make land available for building, consideration for drainage must be made. If, say, a new housing estate of 500 houses is to be built, will the existing sewers be able to carry the added effluent? Will the sewage treatment works be capable of containing and treating the material? All this would be taken into consideration during the planning application to the Local Authority.

Telecoms

Telecommunications or landlines are mainly underground in built-up areas or suspended wires on pole in more rural locations (see also Cooke (2007), p 469). Telecommunications normally run underground in green convoluted conduits. The cabling system can be fibreoptics, blown fibre, standard fibre or where required a bespoke system can be installed. British Telecom can offer a structured cabling system that enables the client to use a single infrastructure that carries both voice and data in any format. The system would have been finalised at the design stage therefore the contractor would only need to agree a connection date with the utility company.

Cable communications

During the design process the client's need for specialist communications would have been noted. If available connection to the cable system would be through a cable service provider. Cable networks are still being ducted in green convoluted pipework throughout Britain. Unlike telephone lines in rural areas, cable communication cannot be suspended on poles therefore the costs of laying and maintaining is offset by the number of potential users.

6.4 Local Authority licences

Licences from the Local Authority to close roads, erect hoarding with areas for advertising, close footpaths or suspend parking bays must all be applied for. Some applications can take weeks to be confirmed.

6.5 Informing the HSE (Health and Safety Executive)

It is good policy to inform the HSE of your intention to proceed on site as soon as possible, although they must be advised within six weeks of starting. Apart from domestic works, it is easier to say that all other work requires an F10 form to be completed for construction projects likely to last longer than 30 days or involve more than 500 person-days of construction. A person-day means a worker carrying out one shift per day, therefore 50 persons on site for 10 days equals 500 person-days. The form F10 can be viewed on the HSE website: https://www.hse.gov.uk/forms/notification/f10.pdf.

The form indicates the range of works that should be notified to the HSE. There are exceptions, so a phone call for advice would be useful. The word 'domestic', for example, means the work would be small in nature such as a house extension or a conservatory. A new build house or houses and flats are notifiable under: New Build – Residential Premises. The person or persons who should notify the HSE are the client and/or the CDM co-ordinator. Information about the project, expected duration, type of work, name and addresses of the main contractor etc. enables the HSE to prioritise which sites need the most attention. Contractors that have a poor record of health and safety issues obviously require more monitoring than those who have exemplary record.

6.6 Suppliers

Suppliers will need to be appointed. If the contract is in an area where the contractor has not previously worked, then accounts with some local suppliers will be required. National chains are less of a problem as the account facility can be transferred to the local branch, whereas other suppliers will need to open a credit account.

Credit accounts are designed to enable a contractor to order materials and pay for them at the end of the month. If, when the statement of account is sent through, the contractor settles (pays the account), the credit limit returns to the full agreed amount. If, however, the contractor does not settle the account, the amount of materials that can be ordered will be restricted to the credit limit. Many credit accounts allow a discount if they are settled by the required time, otherwise the original sum will be due. An estimator may have worked on the figure based on the cost of the material plus profit. If the contractor does not settle on time the difference could be the standard $2^1/_2$%, which will reduce the profit the contractor had planned to make. In circumstances where the contractor has not settled the account the supplier will stop supplies. The consequence could be that the job will come to a standstill until new supplies can be found.

Contractors rely on cash flow, even the really large companies. The bigger the contract, the more the materials will cost, therefore a supplier will want some assurance that if they supply the contractor they will be paid. To set-up an account the contractor normally has to complete an application form providing details of their company such as the trading name and address, names of the directors and the bank they use. The supplier will apply to a credit reference agency to check that the company is solvent and has no legal cases pending against them. The agency will provide a suggested credit limit based on the liquidity of the company and its assets. A company with a short trading record or very small turnover will appear as a risk therefore the supplier may not offer a credit account until a credit history has been established. In the past suppliers would ask the contractor for two referees who could confirm that they have supplied materials to the contractor without any problems. Now business in the supply industry is becoming very difficult; especially with the Data Protection Act suppliers are reluctant to give any data they hold other than to official companies.

6.7 Planning

Planning the delivery of plant, materials and labour is a very complex task. Under the heading 'procurement' planners will pre-programme the whole site operation. Before that takes place a project method would be agreed, usually by the contracts manager, project manager and senior planners. During the tendering process a plan of operation would have been proposed outlining the method of construction.

If the contract is a standard building contract (SBC) the tender documents may include:

- a bill of quantities
- drawings (termed plans)
- specifications.

The contractor will have to decide the type and number of plant and the labour required. For example, would a tower crane be more useful than a mobile crane? Would buying the crane, leasing the crane or renting the crane be the most cost-effective? Is it better to use concrete skips in conjunction with a tower crane or a static concrete pump adjustable as the work proceeds? In contrast, a mobile concrete pump may be more costly per cubic metre to place the concrete yet work out cheaper overall as the plant will only be on site when required.

6.8 Bar charts

When the overall plan has been agreed and the tender has been won the management team will review the plan again to ensure it is still the most appropriate method. The contractor will have to supply the operational plan to the CDM co-ordinator to ensure that the proposed methods are acceptable and safe. From the agreement the planners will take over. On a small project a Gantt chart (more commonly known as a bar chart) would be drawn up (see Figure 6.1). A list of the various stages or operations of the project would be written in job order on the left hand column. To identify the stages numbers are prefixed to the left. At the top of the chart from left to right the timescale is written in. The start and finishing dates are part of the contract therefore the planners have to fit all of the stages within the time parameters.

Henry Laurence Gantt modified the bar chart in the early 1900s to suit construction projects, although the theory remained the same. His innovation was to show the timescale on the 'x' axis against the stage or item on the 'y' axis.

	Week no.	1	2	3	4	5	6	7	8	9	10	11
	Operation											
1	Set-up site											
2	Excavations											
3	Concrete											
4	Brickwork											
5	G/Floor											
6	Brickwork											
7	Roofs											

Figure 6.1 A Gantt or bar chart.

Bar charts up until then had not included the time element. One of the early large projects was the Hoover Dam in 1931, followed by the Interstate Highway projects in 1956. The left hand column can be a series of grouped stages to reduce their number or itemised in greater detail, making the chart more complex and larger. For example, a Gantt chart for a fire station covered an area about 2.00 m × 1.5 m wide on the site manager's wall.

Additional Gantt charts can be produced to programme the materials, plant and labour requirements. In a similar way, the stages are written in a column to the left, while the bars indicate the timing and duration of the plant requirements, when materials are required and the amount of labour required. If everything goes according to the plan the Gantt chart should be adequate for all of the project team to work to. The project manager will know what is required, when it is required and who is required. He or she can fine tune the on-site plan knowing it should enable completion by a set time. However, there are several non-programmable factors:

- weather
- machine breakdown
- delivery problems
- labour problems
- accidents.

Starting with weather, the planners would take the time of year into account when planning the stages. For example, excavating for foundations in January will be different to, say, July. In winter there are less daylight hours for working, the temperatures are likely to be cold and there is a possibility of snow. In contrast, the summer months will have longer daylight hours, allowing possible longer working hours. The higher temperatures are especially important where concrete and masonry are being laid. At early and late times of the year high winds tend to be more common therefore time out for crane operations will be higher than in the summer. In recent decades the weather patterns are less predictable. In the summer of 2007 there were extensive floods in the south and midlands of England that could not be foreseen. In October 1987 hurricane force winds battered the south of England and again in September 2006 with wind speeds of 75 mph.

The term used for weather that prevents operations is inclement weather. Generally the weather needs to be unusual for that time of year if a claim for delay is to be made. As previously mentioned, snow in winter is to be expected. In recent years the winters have been warmer than usual however warm winters cannot be classed as normal therefore unless the snow is extreme it should be taken into account.

Example

A contractor has to plan for the construction of a pair of houses. The commencement date is the first week in May and completion in the second week in July, 11 weeks in total.

The list of operations:

- make site secure
- clear site and scrape work area of topsoil

Continued

- mark out foundations
- mark out service ducts
- excavate foundations
- make provisions for service ducts and drainage
- pour foundations
- install all service ducts and drainage connections
- bring up substructure wall to dpc level
- apply weed killer to beneath floor area
- insulate water mains ducting
- lay dpcs
- place block and beam floor.

As you can see, if every operation is listed, the column is going to be very long indeed, therefore the operations are grouped together, thus:

- 'Site set-up': could be used to include making the site secure, removal of topsoil from the working area and storing on site, marking out for foundations, service ducts and drainage below the building and through/beneath the foundations.
- Excavations: for foundations including concrete.
- Ground works: referring to placing service ducts and drainage pipework that will be beneath the building and go through or below the foundations, and pouring concrete into the excavations.
- Brickwork: would cover all masonry from foundation level to finished plate level, innerleaf blockwork, cavity insulation, provision for window and door openings, provision for first floor joists and fixing of the wall plate. It would also include internal loadbearing walls at ground floor level and non-loadbearing partition walls at ground level. If the design has incorporated non-loadbearing blockwork partitions at first floor they would all be included under the heading Brickwork.

The list would continue with all stages of the contract usually finishing with landscaping and hand over. By using generic headings for the type of work the site manager can more easily see what plant, materials and labour will be required and for what duration. Materials could be planned against a 'call off' from the supplier. A 'call off' is a term used when an order is placed with a merchant or supplier for a total quantity of a specified material to complete the job and then taken in small batches. For example, say the two houses will require 9000 facing bricks plus 200 m^2 of 100 mm concrete blocks. If they were all delivered in one go it is unlikely that there would be sufficient room on site for storage. Also the contractor would have to pay for all the materials whilst they were being stored onsite. With a 'call off' the site manager can plan smaller mixed loads of bricks and blocks to keep the brick layers in materials. Extending the delivery period over two months would also spread the cost of materials outlay – see Figure 6.1. The disadvantage with calling off small loads can be that it is more costly than full loads from the manufacturer (less handling costs taking a works delivery than materials from a merchant's yard) and colour variations of facing materials. A good tradesman and labourer will use two or more brick packs at the same time, thus mixing the bricks from the different packs. Where mixing does not take place, lines or patches of colour variations will be noticeable forever (see Figure 6.2). The architect and/or client are well within their rights to have the unmixed brickwork taken down and replaced at the contractor's expense and with no extension of contracted time.

The bar chart/Gantt chart has limitations. If the programme goes according to the plan everything should be complete by week 11, ready for handover to the client. However, what happens if the bricklayer gang is held up on their previous contract? Will they stop working and not complete the work so they

Figure 6.2 A poorly mixed batch of facing bricks.

can start on time with your contract? If the bricklayers are part of a firm of bricklayer sub-contractors they may be able to pull a bricklaying gang (commonly two bricklayers and one labourer) from another job to make a start on your contract and then work a weekend or longer hours to catch the time up. Alternatively, the brick lorry may have mechanical problems therefore the delivery is late and the brick layers cannot proceed with their work through a lack of materials. In these circumstances the programme will need to be rearranged.

The start and completion date should remain the same, though some operations may be able to proceed in a different order. On the pair of houses it is very unlikely that anything could be changed at the beginning. Clearing the site area, setting out, excavations, pouring concrete etc. all need to be carried out in that order. However, as the work proceeds depending upon the problem, for example the late delivery of roof tiles. it may be possible to sark out the roof enabling the block layers to build non-loadbearing first floor partitions and tackers to fix plasterboard ceilings. (Tackers traditionally fix plasterboard sheets to the underside of floor joists and ceiling joists ready for the plasterers to scrim, joint and skim. Sarking is an underlay material used to provide a second layer beneath the roof tiles to prevent fine snow or wind blowing into the roof void.) A common site term would be 'the building is weathered', meaning the roof may not be finished but internal work can proceed as the rain, wind and snow will be kept out by the sarking materials. For the building to be weathered the windows and door openings will need to be sealed off. If the windows still require glazing, sheet polythene can be used to prevent the rain, wind and snow entering the building. If expensive hardwood doors or Upvc doors have been specified, then temporary cheap door blanks or shuttering ply doors make ideal substitutes until near the handover date, thus preventing damage by the various trades.

6.9 Programme Management Software

There are several software packages available that are Windows-based, working on standard PCs. Microsoft Project is one such package. An alternative, and in my opinion a more useful, tool is Powerproject from Asta Development. To see a working demonstration, access their website on: www.astadev.com/software/powerproject/democentre/index.asp. See also Chapter 10 for comment on a real project.

So far we have looked at:

- securing the site
- procurement and setting up credit accounts
- planning the various activities.

6.10 Procurement of materials

The designer has produced a set of working drawings plus a schedule of works to be carried out. Depending upon the type of contract, a bill of quantities may have been prepared by the PQS. The successful contractor has tendered for the work and won the contract. The estimators and planners have agreed a plan of action and had it approved by the CDM co-ordinator.

A plan of work has been produced by the planners, who will make arrangements with the various suppliers and sub-contractors. Quotes would have been gathered at the time of tendering therefore confirmation of acceptance will be required. After a contract has been won the contractor will spend more time working out detailed plans for the project. Planning is very costly and time-consuming therefore at pre-tender stage minimal work is carried out for the cover price. i.e. the figures put in to the tender should be adequate to cover the materials, labour and plant plus overheads and profit. When the contract has been won the planners will try to improve on the costs and produce greater detailed plans.

Where a contract has a bill of quantities the CQS (contractor's quantity surveyor) or cost consultant will dissect the sums allowed for in the tender into a series of budgets. Setting up site, hoarding, security, site accommodation, temporary services, office equipment etc. will be covered in a lump sum, commonly as part of the preliminaries. The bulk of the work will be contained as 'measured work', grouped in trades as per the SMM7 categories.

Prime cost (PC) sums are lump sums that will cover work carried out by specialist nominated sub-contractors. For example. a specialist glass and glazing sub-contractor who has been nominated to supply and fix for the sum of £XXX. The main contractor would have had a quotation from the sub-contractor at pre-tender stage and then added an extra sum for attendance. This allows the main contractor to put an additional charge for the facilities that will be available on site. The extra sum for attendance will be part of the preliminaries sum.

Provisional sums (PS) will be shown in the BoQ as a lump sum designed to cover the cost of work that could not be priced prior to the billing stage. For example, the client may not have decided on the type of brick or roofing tile before the PQS set out the BoQ, therefore a PS would be entered which should cover the cost of whichever brick the client eventually selects.

Planning would be carried out by the site manager with help from the estimator and buyer in small companies. On large contracts the planning team would comprise:

- planners
- contractor's quantity surveyor or cost consultant
- site managers – on large sites the project would be divided up into sections or areas and allocated to a several site managers. Each site manager can then concentrate on their specific area
- buyers – will need to be in constant contact with the planners to ensure materials or plant that require a long lead in time (time required between placing the order and supplying on site) are ordered on time
- estimators – although their input is mainly pre-tender stage, the planners will need to be able to discuss how the estimators obtained prices and where from. It is usual practice for the estimators to pass over their files to the contract team. However, personal information is very useful when trying to interpret the paperwork.

6.11 Plant

When the main plan of approach has been settled, team meetings with the site personnel would be arranged. The planners will have to predict the amount, the type and when the plant will be required. The sum set aside for the preliminaries ideally has a contingency sum to allow for extra items. However, if it is not used it will add to the overall profit on the project.

When planning for plant three considerations should have been made:

1. Is it worth buying the plant?
2. Is it worth hiring the plant?
3. Is it worth leasing the plant?

Consider a 360° excavator. It is a very expensive piece of plant to buy outright unless it is going to be continually used. Buying the plant new can be advantageous as it can be set off against the profits of the company over several years, thereby reducing the tax liability. A company of groundworkers would use that type of plant continually, therefore it would be worth buying it. Buying the machine would also mean 'down time' whilst it is serviced plus the servicing costs. Companies that have several pieces of heavy plant may find it cost-effective to have their own fitters and service engineers to maintain the plant.

In contrast, a building contractor may only require the machine to carry out low level demolition, and groundworks, therefore it would be better to hire the machine with or without a driver for the short period required. The cost per hour will be greater than owning or hiring, but it is offset against not having the servicing costs, down time for servicing when the machine cannot operate and most of all the depreciation on capital outlay required to buy the machine.

Where a contract requires the use of plant over an extended period, hiring may be cost-prohibitive based on a daily, weekly or monthly rate. Buying the plant will require a massive initial outlay or a capital loan and interest costs. Leasing in those circumstances may be the best option. Leasing the machine for, say, a year and renewal is required for a second project can work out cheaper than hiring without the initial capital costs of buying. Servicing costs are normally part of the leasing agreement. Some leases have an option to buy at the end of the lease at a reduced market price. The capital outlay is then spread over the duration of the lease.

The same reasoning can be said of many pieces of plant. Haulage and storage are also factors when considering plant. A dozer will require a low loader lorry to transport it from site to the storage area or next site therefore the costs of the specialist lorry must be considered.

When the planners select the method of payment for the plant cash flow has to be taken into account. If, say, the financial outlay for a machine is £70 000 and the contractor has calculated that the profit on the project is enough to buy the plant, the profit will only be available at the completion of the job. Taking a loan out based on paying it off when the job is complete may be scuppered if the costs run over budget. The loan will also be set against the amount the company can borrow, which can also effectively reduce the buying limits on credit accounts.

Selecting plant is based on use and output. The size of an excavator will be based on the volume of the dig, the depth of the dig and the materials handling. A 360° excavator with an extending boom may be able to excavate to a greater depth, albeit more slowly than a face shovel. The face shovel could be used to load tipper trucks though. The planners would either:

1 decide to sub-contract the work out to a specialist company who would supply the plant and labour to carry out the whole operation
2 calculate how fast (output) a machine can excavate a volume of soil and whether it is better to have the same machine load the muck clearance vehicles or to have a separate loading shovel. If there is a large volume of spoil to be removed, the excavator would be better utilised digging the material out and a second machine, the loading shovel, moving the spoil into heaps ready for loading the muck clearance lorries. Would it be more effective to have two loading shovels, one to move and stack whilst the other loads? The time factor will influence the decision. Is it better to spread the operation over a short period and use more plant or extend the activity and only have minimal plant? If a BoQ has been used, the description should indicate the volume of spoil and the distance a spoil heap will be from the dig. The planners will be able to base the size, number and type of plant on that information.

6.12 Site production

When the site is operational, the following processes will take place:

- Plant, materials and labour will be shown on a chart or programme – see bar charts and Gantt charts.
- The site will place a requisition to the main office buyers to place official orders for the plant, materials and labour.
- The site will then call off as required either directly to the supplier quoting an order number or via the buying office. It is important to have a job number, contract number or order to tie the requisition to the order for charging purposes. A company may have more than one job running at one time therefore it is essential to know which job the order relates to.
- When the order is delivered or collected it should be signed for and dated. The company will be invoiced at the end of the calendar month for materials or plant. If there is no record of receipt it is very difficult to account for the invoices. For example if a delivery of blocks is made to site the lorry driver will have delivery tickets. The ticket will show the number of blocks or parcels of blocks that are being delivered. The driver will require a signature from the contractor if the vehicle is to leave the public highway, to state that any damage caused will be settled by the contractor. An example of damage could be a lorry driver being asked to drive off of the roadway to be unloaded by the site fork lifts. The lorry weighed in excess of 30 tonnes and the wheels sank into the soft clay up to the axles differentials. When the lorry was unloaded it could not drive off so the contractor used a dozer to push the tailboard of the lorry to help it out of the mud. The consequence was the flat bed of the lorry buckled and caused several thousands of pound worth of damage. The lorry had to have a new flat bed and was off the road for several weeks. If possible do not take heavy goods vehicles off the road or hardstanding. When the delivery has been made the driver will require a signature to state the goods have been received in good condition. Again it is important as the materials may be damaged by 'an other' after the delivery has been made. A lorry load of blocks had been delivered and set down as instructed by the site personnel. Shortly after another activity was to be carried out and the blocks were in the way. The contractor used a multipurpose excavator to push the two high stacked parcels out of the way. They later claimed the blocks had been delivered in a damaged state.
- In normal circumstances the materials are delivered in a professional manner and signed for. The delivery driver retains a copy of the signed ticket and returns it to the office of the supplier. The ticket is then booked in and sent to the invoicing department to be added onto the contractor's account. At the end of the month the supplier will send a copy of the invoice for payment by the contractor.
- On site the driver would leave another copy of the signed delivery ticket as a receipt for the goods. Many companies use a 'two part note' which is self-carbonising, ensuring both parties have a duplicate of the

document. Some companies also have a computerised delivery, where the signature is written on an electronic screen and a paper receipt is issued. Either way, the site has a record of the goods that have been delivered. The delivery ticket should be logged in a site diary recording materials movement and then sent to the buyers to be checked off against the order. If it is confirmed as part or the full order, it will be passed to the accounts department to be checked off against a specific order and job number. The materials will then be booked against a specific job. When the invoice from the supplier arrives at the end of the month, each delivery ticket will be checked off to ensure it matched what had been delivered. Palletised materials often have a deposit charged on the delivery. The pallets when returned to the supplier in serviceable condition will be credited against the credit account. The accounts department will need a copy of the receipt for returned pallets to deduct from the invoice. As you can see, if all the people involved do their part of the paperwork for deliveries then the payment should be straightforward. However, if the paperwork is slow or lost, the knock-on effect can have high financial consequences and could lead to an account being 'put on hold', i.e. no further deliveries to be made until the account has been cleared/paid in full. Contractors could be starved of materials, all because someone has not paid the invoice. The consequence would be to put the contract behind schedule.

So far we have looked at:

- pre-production planning
- planning plant requirements – comparing the advantages of buying, leasing and hiring. Considerations of number, size and type of plant
- the roles of the planners, estimators, buyers, site management and accounts department
- producing budgets from a bill of quantities
- the basics of setting up credit accounts
- the function of a delivery ticket.

When the site is operational the management control the various activities based on the production plans. The sub-contractors normally have their own management on site or at least in daily contact. For example, a sub-contracting bricklaying company may have a contracts manager who controls several teams of bricklayers on several sites. He or she would liaise with the site management and attend site meetings but leave the day to day running of the work to a bricklaying foreman on site. The trade foreman would be paid slightly more than the bricklayers for the extra responsibility. He or she would be responsible for ensuring toolbox talks are carried out and that all employees of the firm receive the correct personal protective equipment (PPE) prior to working. Other responsibilities include taking instruction from the main contractor's management team. For example, on larger contracts specific areas or parts of the project would be allocated to specific site managers, who would organise their part of the contract. The site manager would ensure the materials are on site at the required time and in sufficient quantity. The trade foremen would be given instruction by the site manager where the work is to be carried out and a set of drawings showing dimensions, bond of the masonry, and the required finish such as 'face work' where the brick, blocks, or stone would be left exposed. The trade foreman would keep the site manager informed of the on-site stock of materials and any scaffolding requirements for the near future.

6.13 Quality of materials and workmanship

The quality of workmanship and materials should be set out by the designer at the design stage, well before the work commences. Samples of the intended materials would be obtained to enable the feel and true texture to be experienced. Coloured photographs in a brochure will offer an idea of the general aesthetics of the material; only a sample will show the true colours. The samples of the finishings such as wall coverings, paint, floor and ceiling coverings etc. are normally assembled on sample boards to enable all the associated areas to be compared together.

Where a material is to be assembled from component parts such as brickwork a brick sample would be obtained from a manufacturer or supplier comprising three typical bricks, two of which will be the extremes of colour variation plus a third typical brick. The designer will show the sample to client and seek approval. Before the bricklaying starts on site the brickwork contractor will build the sample wall using bricks from a standard parcel. As previously mentioned, it is essential that the bricks are not selected more than as normal for the whole contract. If the bricks are easily chipped then they should be used and not set aside. The client and lead consultant need a true representation of the masonry that will be used.

As the work proceeds, the standard of workmanship and quality of materials would be compared with a sample panel. The brickwork contractor would normally erect a sample panel of masonry about 1 m^2 to enable the client and lead consultant to see what the walling will look like when complete. The brickwork contractor should not try to make the panel perfect as the lead consultant will require all of the masonry to be of the same standard. The sample should be of the standard that the bricklayer will be able to maintain throughout the contract. If the standard is not acceptable, then the lead consultant and the client can decide whether it is the material that is at fault or poor workmanship. If it is the latter, then the contractor will have the opportunity of putting better tradesman on the contract who will build another sample panel, or the brickwork contractor will be dismissed.

Traditionally the clerk of works is normally an ex-tradesman with a brickwork or carpentry background. However, in today's environment modern clerks of works will often have a surveying qualification and no trade experience. Many large projects based on Client Management Contracts, such as the Ropemaker site discussed in Chapter 10, do not have clerks of works as there are numerous checks carried out by the site management from the principal contractor and main contractor. Clerks of works are more likely to be used on the larger sites using the traditional forms of contract where a main contractor is contracted via a lead consultant. Local Authority, Government and Public works commonly have a 'clerk of works' on site. His or her function is to continually monitor the progress of the work with emphasis on compliance to the methods of construction set out in the British Standards, Euro Codes and Codes of Practice.

Where clerks of works are employed, their function is basically to check on the quality of the work being carried out on behalf of the architect and client on site. They do not have the authority to instruct the main contractor or subcontractors, it is more an advisory capacity. However, if the contractors do not pay heed to the advice, the clerk of works can advise the lead consultant, who has the authority to instruct the contractor to remedy the situation, normally in the form of an AI (architect's instruction). The type of contract will indicate the authority of the various construction team members. For example, on a standard building contract (SBC) the lead consultant is likely to be an architect or surveyor. They will be contracted to either oversee the running of the project or to simply prepare the contract up to and including all necessary legislative permission only. If it is the latter then they have no authority over the project. Indeed, they will not be paid or insured and would therefore be very unwise to become involved.

Quality can be associated with waste. To achieve an acceptable level of quality many materials will be set aside as being imperfect. Mass production of materials has enabled the costs per unit to be reduced significantly. Materials handling and site storage improvements have further reduced waste. For example, concrete blocks are stacked and banded (steel or plastic bands about 20 mm wide that tightly hold a layer or bundle of materials together) at the last stage of manufacture. Many manufacturers shrink wrap small stacks of bricks or blocks known as 'parcels'. Parcels can be on a pallet to enable mechanical handling by fork lift vehicles or grabs (see Figure 6.3). The advantages of the banding are that a large number of bricks or blocks can be moved at one time by machine and there should be less damage during transit. The grab operator must position the grab in the correct place otherwise the enormous hydraulic pressure required to lift the materials could also crack them. Only a skilled and trained operator should use the equipment. It can be a problem when a relief haulage driver who has the necessary HGV driving licence but no grab experience crushes virtually every parcel of blocks he delivered.

Face blockwork requires high quality materials as the finish will be 'as laid', i.e. no plaster, paint or other covering material. High density blocks such as the Forticrete texture solid block are made from dense aggregates in steel moulds. The finish is considered the Rolls-Royce of concrete blocks. Being so dense and formed in individual steel moulds, each block should arrive on site in perfect condition. If handled correctly the waste will amount to cutting the blocks for bonding and coursing and therefore should be minimal. The designer will probably see a small sample brickette for purposes of checking colour and texture and may not be aware of the physical weight (mass × gravity) of each block and therefore the manual handling required by the blocklayer. A standard 215 mm thick block has a dry weight of 42.7kg. The site management

Figure 6.3 Kelly demonstrating a remote controlled HIAB brick grab.

would require a method statement from the blocklayer sub-contractor stating how the blocks will be laid.

In October 2007 the Manual Handling Operations Regulations 1992 (MHOR) was amended. There is now no set specific requirement of weight limits. Instead the employer will have to assess the risk of injury and where possible remove the risk or minimise it as far as practical. Large companies will no doubt comply with this. However, smaller or less scrupulous firms will continue the old bad practices that have left many brick and block layers with permanent back and joint problems. It is the duty of the employer to ensure that as far as reasonably practical the employee is not involved with any risk of injury whilst manually lifting (see Figure 6.4). The Health and Safety Executive have produced advice known as the 'MAC assessment' (Manual handling Assessment Charts), which although it does not cover all eventualities does give very practical and clear advice on assessment. The method statement from the trade contractor should include a risk assessment of the procedures that will be carried out during the work. In addition, a more detailed risk assessment relating to any manual handling tasks should be included, based on a colour code:

- *green:* low risk more applicable to pregnant women and young people
- *amber:* medium risk and the tasks should be examined more closely
- *red:* high level of risk, action required
- *purple:* very high level of risk that may result in injury.

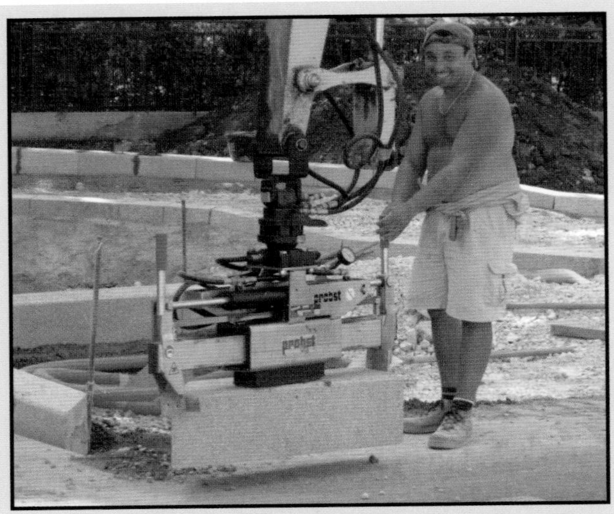

Figure 6.4 Mechanised lifting equipment.

The MAC can be viewed on the HSE website at www.hse.gov.uk/pubns/indg383.pdf.

The employee also has a duty to comply with the employer's safety requirements – for example, to use the handling aids. In practice an employer may supply specialist lifting equipment but the operatives may feel it is 'sissy' and not macho to use them. On the Continent the use of machinery to carry out many building operations has been standard for decades (Figure 6.4). Plastering, for example, is commonly carried out by mixing the powdered plaster with water outside the building using a powered mixer towed behind a van. The plaster is then pumped into the building via a hose and sprayed onto the walls. The finishing is carried out using large floats used by skilled tradesmen.

In contrast, in the UK most plaster is manually offloaded and mixed in a plasterer's bath using a paddle mixer and a power drill. The plaster is then lifted manually into a bucket and placed on a spot board ready for the plasterer to transfer it onto his or her hawk prior to spreading it on the wall or ceiling. The mixing process by necessity has to be close to the operation due to the plaster setting time therefore the dust caused when opening the plaster bags especially when mixing inside the building is another health hazard. Gypsum plasters are basically a form of calcium sulphate and as such is not harmful, but as dust it should not be inhaled. This is another issue of hazard identification and the operatives not wanting to reduce the hazard. Have you ever seen a plasterer's labourer wearing a dust mask? From a practical point of view wearing a suitable mask to prevent the particulate size from being inhaled would be very uncomfortable. The outcome is that the old traditional methods are still being used even though better methods as far as health and safety are available.

Organisation and communication is very important to prevent a shortage of materials, labour or plant. It is pointless having bricklayers and materials standing idle due to the scaffolding not being available. The organisation is carried out on large sites with a team briefing every morning between representatives from the trade contractors and site management. The objectives are:

- to discuss the work plan for the day
- to discuss the work plan for the following day
- to update on the work from the previous day.

Any issues that any of the trade contractors have can be brought to the attention of the site management. For example, noise, dust or fumes may be affecting another trade working in the same area. Where possible the issues should have been noticed before the work was carried out and planned accordingly. In practice, though, materials may not fit, for instance, and therefore need cutting to size; and so an issue arises. Any 'hot work permits' requests would be made for work being carried out the following day. This enables the site managers to check other work being carried out in that area and check the method statement from the trade contractor. In addition to the daily meetings, weekly meetings are held between the site managers and senior project management. The client's representative and the lead consultant are normally at the weekly meeting to discuss the progression of the works. Many sites have a monthly site meeting where the client, lead consultant, cost consultant, project management, site management, and trade contractor's management review the progression of the works.

Chapter 7

Acts and regulations

Many references are made to laws and Acts, but what actually are they? This section of the book gives a brief history of how they came about, the differences between Statutory Instruments and Regulations, and why they evolved. The study of history should be the process of learning. As a child, if we touch something that is hot we hopefully will learn from the experience and have the knowledge for the future – don't touch that as it is hot. Education perhaps? If we look at the events of history we should again learn and remember the outcomes and consequences.

Something as important as dumping raw sewage in the River Thames gave rise to the 'Great Stink'. Today, in October 2008 London is once again dumping thousands of cubic metres of effluent in the same river. The reason is the same; more population than the sewage system is capable of handling. There are many important people wanting to rectify the matter (the same as in 1858), but who will pay for the work to be carried out?

Some of the myths of history have also been discredited. Thomas Crapper – did he invent the flushing loo? Why did the Building Act come about? What are the Approved Documents? – read on.

Part of being a member of a society is to obey the rules. Without rules society could not exist. Even in a dictatorship everyone is expected to obey the rules. Historically the leader of a group or tribe would have several elders to help organise everyone. Eventually the rules would be written down either as religious texts or legal texts. Today in Britain the rules are still written down and to give them the authoritative power they are signed by the Sovereign.

7.1 Who actually puts the rules together in the first place and why?

Originally the sovereign would have his or her council, a group of influential people including religious leaders, who made the rules. Wealthy people would try to gain favour by offering money (bribery) to change or add rules to work in their favour. The most important rules became laws and written on documents, the most important of which were related to ownership of property and chattels (moveable items belonging to a person). Basically wealth required protection therefore many of the laws are related to taxation (wealth to the sovereign) and in favour of those with wealth. In 1215 King John signed the Magna Carta, restricting the power of the sovereign, the turning point in law-making in Britain. By 1295 King Edward I assembled representatives from the population (all wealthy influential people), the start of 'Parliament'. The rules and laws now were on behalf of the population, acting on their behalf, hence the title of 'Acts of Parliament'. In 1341 the house divided into two sections, the lords and the commoners. The title 'Lord' has its origins from William I in 1066. Total sovereign of England he needed trustworthy people to control the conquered nation. The land area was divided up and placed under the control of his noblemen, earls and barons as the introduction of the feudal system. They were peers of the realm equal (in theory) to the realm. They had the authority to preside over county courts and pass sentence and even pardon murders and those accused of treason.

Feudal systems are based on the landowner allowing a person or family to control an area of land in return for a sworn oath of allegiance to the king. In this case William I took possession of England and shortly after, to take stock of what he had acquired, instigated a national audit recorded in two volumes, known as the Domesday book.

The new controllers were termed 'tenants-in-chief'. Basically they were the head person who could live off the land allocated to them as tenants of the king. (William 1 was the supreme landlord). The areas of land were vast and later by Medieval times became counties and shires. As part of the deal they had to collect taxes for the crown and provide knights who would fight for and protect the king. Therefore to reward the knights (soldiers on horseback) they were allocated land under the feudal system. The knights became under-tenants and controlled smaller parts of land known as manors. A manor commonly comprised a village, church, three large arable fields and the home for the knight and his family, the manor house. They became lord of the manor. As lord they basically ran the manor, taxing the peasants, who worked small strips of land in the fields. The lord administered the law and held court to preside over peasant's disputes. More serious cases would be referred to the county court. Knighted by the king, lords had special privileges that could be passed down through the family on death, hence most of the lords until recently were descendants as opposed to newly appointed. Governments have the authority to apply to the sovereign for permission to elevate a commoner to become a lord. The Labour party ended the right that enables peers to be able to pass down their rights through the generations under the House of Lords Reform 1999.

What are the Houses of Parliament?

The Houses of Parliament comprise the Lords and the Commons. The former have traditionally more privileges and are termed the 'upper house'. Currently 750 members including Life Peers (the peerage ceases on death and cannot be passed on to their children), Law Lords, bishops and elected hereditary Peers (the 1999 House of Lords Reform stopped the right of hereditary peers to sit and vote in the house, however, 92 members were elected by the Government to remain until the next part of the reform process). An excellent website to visit to learn more is www.parliament.uk. The commons are people who have been elected by the population to act on their behalf from the constituencies. In the 18th century the members of parliament tended to form two main groups, the Whigs and the Tories. The Tories tended to be landowners (a title by deed) and the Whigs tended to be educated professionals, later to become Liberals. The Labour Party formed in the early 20th century as a breakaway part of the Liberal Party mainly funded and influenced by the Trade Unions. When a group/party have a significant number of MPs the sovereign will ask whether they will form a Government on the sovereign's behalf. The important factor is that the sovereign is still the head of the country, which is termed a sovereign state. Motions for laws or private members bills are put forward by MPs and discussed in the House and then referred to the Upper House via the Law Commission. If and when agreed the Prime Minister will ask the Queen (sovereign and head of state) for permission to enter the law onto the statute book. Originally all laws were written in a book. However, with the vast number of laws passed over the centuries, the laws have been written on documents and stored in a vast room. The Law Commission has the task of reviewing the laws and where appropriate requesting repeals to have them removed from the statute book. The repeals have to go through Parliament and have the full permission of the sovereign.

Now we have looked at a potted history of Acts of Parliament and why County Councils have authority, we can look at the development of the Building Act 1984.

7.2 Health and environmental laws from William I to Charles II

Generally the Great Fire of London is considered as the beginning of building regulation, but many regulations date back centuries before 1666. The most poignant of regulations came with William I relating to fire. Common in Europe the population of walled towns would have to extinguish or cover open fires. A bell would toll indicating anyone outside the walls to return before the gates were closed and the curfew began. In the early 11th century William decreed a curfew within the City of London each night. (The word 'curfew' means 'cover fire'.)

Later in the 15th and 16th centuries coal became a popular fuel as wood became scarce. Ships brought coal from Newcastle down to London and Bristol. The Tudor period saw the introduction of the chimney. Before the new design, buildings had a hole in the roof to enable the smoke to escape as burning wood if dry gave off little smoke or noxious gases. In contrast coal, especially sea coal, gave off thick black sooty smoke when burned. However, the cost of wood between 1450 and 1650 rose by about 400% – and by about 780% in London according to Brimblecombe's book *The Big Smoke*. Even in the 13th century they had known that smoke was bad for the health of people and animals and in particular livestock. King John in 1215 had tried to curb the use of coal. At that time the coal was known as 'sea coal'. As the cliffs eroded into the sea the coal seams broke into lumps that floated on the water. Locals would gather the sea coal when it washed ashore.

Other regulations in the 13th century made by the City of London required every man to keep clean the front of his own tenement. This was followed in 1309 by regulation prohibiting the casting of filth from houses onto the streets and lanes in the City. By 1357 a Royal Order stated that no rubbish or filth be thrown into the River Thames or Flete (the spelling changed to Fleet some years later).

During the 14th century Edward I's proclamation banned the use of sea coal owing to the high volume of smoke given off. During the reign of Elizabeth I (1558–1603) 50 000 tons of coal went into the city of London each year. John Evelyn, founder member of The Royal Society, in 1660 noted that coal burning was creating deleterious health effects and atmospheric pollution. The society even noted the corrosive effects the smoke had on buildings, which we now term acid rain.

Externally, 1562 saw the Highways Act requiring parishioners to spend six days annually to maintain the highways. A Planning Act of 1589 attempted to reduce the building and maintaining of cottages unless there was four acres of land with them. The idea was to stop the population becoming too dense, leading to a shortage of food. Most of the Acts applied to London, though Parish and County Councils followed with bye-laws and their own Acts.

During the 14th and 15th centuries plagues and fires had destroyed much of London. There were five major fires during the early part of the 17th century. James I in the early 1600s declared, 'that he had found a City of timber and wanted to leave a City of brick'. However, apart from a few buildings in the new Covent Garden area, most new buildings in London continued to be built of timber. Rules made as bye-laws prevented the use of thatched roofs on new buildings in London long before the Great Fire of London, though there were many older thatched buildings. In 1666 during a very dry hot summer 13 200 houses and 87 churches became ashes in just five days. The main reason for the fire spreading so quickly was that the area of Pudding Lane was close by to the wharfs and waterfront where tar, pitch and rope together with stacks of timber were stored. The houses were mainly Tudor timber frame with some buildings of brick or stone. The streets were very narrow and most houses and businesses had stacks of timber, some had coal, alcohol and fodder (bales of hay and straw) for their animals. The heat generated from the burning pitch and tar was enough to combust most materials in its wake.

After the Great Fire in September in just three months about 150 new houses had been built from brick or stone. Six commissioners had been given the task to rebuild London. Various designs were put forward based on the cities of Venice and Paris. Streets were to be widened and boulevards created together with squares and grandiose buildings. However, the landlords didn't want to wait. They were losing money from rent. The infamous Bishop of London, Humphrey Henchman, landlord of much of the charred remains around St Paul's Cathedral, continued to charge his tenants rent for buildings that lay as ash. He even went to the extremes of employing bullies to collect the money. Henchman was taken to court under the new Rebuilding Act 1667 that now protected tenants. Prior to the act tenants were responsible for the rent and rebuilding of the premises. Bearing in mind that property insurance did not exist as such on land, the tenant would have to finance the rebuild himself. Although insurance started in the 17th century for the cargo on ships, in particular slaves being transported from Africa to the Americas, there was a type of buildings insurance in London after the Great Fire of 1666.

The birth of modern day insurance

Maritime insurance started when a group of wealthy men who for a premium would underwrite (pay out) for any cargo that did not reach its intended destination. The voyages could take up to 18 months from Europe to Africa then America and back to London, Bristol and Birmingham in England. (The first recorded fraudulent claim was for slaves that were thrown overboard the slave ship Zong. The full history of insurance is out of the scope of this book).

Property insurance came about again by wealthy people starting private fire brigades. Most householders in London were tenants who rented their property, and the few that owned property could for a small weekly sum have a 'fire badge' attached above the front door of their dwelling. The badge would indicate that they paid into the private fire brigade and in the event of a fire would summon assistance. There were several brigades and they would only attend fires that had the bronze badge identifying their company on display. If it wasn't their company's badge they would not help to extinguish the fire. Brigades would be equipped with leather buckets, fire beaters, barge hooks and wrecking bars, some had hand drawn carts to carry water to the scene. As rival fire companies set themselves up, gangs of thugs would be employed to prevent the fire brigades attending the fire, thus only the strong brigades survived, the start of fire insurance to save property.

Why did King Charles II ask Christopher Wren to re-design London? Or did he?

After the Great Fire King Charles II made a Royal proclamation that became the Rebuilding Act 1667. Christopher Wren (childhood friend of the King) and Robert Hooke (an eminent scientist who like Wren became an architect; his most famous structure that is still standing is the Monument) headed the six commissioners/surveyors who were to oversee the rebuilding of London. Wren put forward proposals for the rebuilding of the City influenced from his love of Parisian and Italian town planning. He wanted wide tree-lined streets and squares (boulevards and piazzas) and to open the heart of the city. However, the landowners hated the idea of losing their property and Wren's proposals were halted. He did win contracts for the rebuilding of several major buildings; 51 new city churches and of course St Paul's Cathedral in 1669. He later became knighted in 1673 by King Charles II for his services to the crown.

Many landowners wanted to get their buildings back and businesses working again. They proceeded without waiting for a new master plan. Most of the new houses were based on plans of four Tudor styles used before the Great Fire. The main differences from the Tudor designs were that they were now clad or built in brick and many had the new vertical sliding sash windows from France and Holland.

Soho is a good example where the buildings are still close together, in contrast to the wider streets with additional footpaths in the affluent parts of London such as Lincoln's Inn Fields and Covent Garden. They were laid out in the 1640s and had escaped the Great Fire. Wealthier people living in the heart of London in the mid-1600s used sedan chairs (another influence from the baroque continent) when leaving their homes. The streets were too narrow for horse and carriage and walking in fine clothes would be asking for trouble from the rogues of the time.

In 1680 room heights of new buildings had to conform to new regulations and bye-laws. Up until then new buildings could have any storey height without a minimum. It is a popular myth that the reason for low ceilings in Tudor building was due to the population being much shorter. If buildings reflected the size of the population then why do wealthy people have much larger doors and higher ceilings to their rooms? Economics provides the answer: if the owner is wealthy he or she will want to display such wealth and one such way would be buy longer lengths of timber and straighter lengths of timber. Bricks would be expensive and more so stone, therefore a limestone mansion with vast clear floor areas requiring extraordinary sized timbers and large areas of glazing would indicate a vast fortune. One such building still stands today: Somerset House on the Embankment. Originally built in 1547 for Edward

Seymour, the wealthy uncle to Edward VI, it was not a popular project by any account. His opponents had Seymour arrested for treason and executed in 1552. The grandiose palace became a Royal Palace and home to Elizabeth I. The structure today is the result of a major reconstruction designed by Inigo Jones and parts by Wren. The building is still by today's standards a massive display of wealth and definitely worth a visit.

Enter the 'Industrial Revolution'

The next major event that had an effect on population requiring new Acts and Regulations was the Industrial Revolution. The main development was the use of coal as a fuel. Up until then wood had been the main source of fuel. To make iron extra heat and carbon is required, therefore charcoal had to be produced and transported out of the forests. With no roads charcoal, especially in the wet months, became very difficult to transport, so iron works were normally on a small scale in the forests. Abraham Derby II had developed a way of smelting iron from ore using coal to make large cooking pots. Richard Reynolds, Abraham's son-in-law took over the works from Derby II as Abraham Derby III was only 11 years old at the time. Between the Derbys and the Reynolds iron rails were developed in 1767 and then in the late 1700s large volumes of iron were being transported around the world. The explosion of the age of iron gave birth to larger machines, mills no longer had to be powered by water mills, and steam engines were fitted to ships putting back supremacy of the seas.

The consequence of large volumes of iron and eventually steel meant that wealthy entrepreneurs built massive brick mills and factories on any land close to navigable rivers or supplies of iron ore. The new power houses still needed a water supply for the steam boilers and as a route to move large volumes and weights of materials. Cottage industry and farming had given way to manufacturing on a large scale. To provide the manpower for the new factories and mills, scouts went out into the surrounding countryside offering farm hands a new life in the new towns being built. The population of England had been regulated mainly by disease, and the supply of food and water. Plagues were relatively regular in the main cities such as London, Bristol and Liverpool, brought in from the Far East trading vessels. The problems were mirrored all over trading Europe. In contrast, people who lived in the towns and villages still experienced the plagues albeit to a lesser extent.

Living in a village reflected the wealthy and the poor. The wealthy could afford to live a luxurious life wanting for little. Clothes for the rich would come from the fashion houses of Milan, and Paris with fine silks from the Far East. Wool and cotton woven into cloths traded across Europe. Large estates ensured an income from the workers on the land. Large numbers of people were required to tend the crops or look after the livestock. Trading the grains, vegetables and wool required markets. With no hard road surfaces or refrigeration vehicles, meat tended to be driven on hoof and slaughtered on site.

As previously mentioned, most of the laws were aimed at protecting possessions. A farmer or even a person with a few animals to sell would require paperwork as evidence of ownership. The peasants rarely had enough surplus to sell therefore would not go to market. What little meat or crops they had would come from the farmer or landowner as part of their pay. Their homes were commonly tied to the estate. meaning the landowner owned the home and could literally throw the occupants out at any time. The quality of building varied from estate to estate. Water generally was via a hand pump or bucket and well. Sanitation was basically a hole or wet pit where effluent would be thrown and when the pit was full another pit dug. Ground floors were commonly battered mud or some of the better cottages had flagstone floors. The cottages may have had a small fireplace with a larder (very small room to store food) either on the ground floor or upper floor next to the chimney. The idea was to use the heat from the chimney to keep the food dry.

Meat such as rabbit, hare, pigeon etc. or for special occasion pig or venison (typically poached from the landlord's estate) would be hung then salted and stored in the larder. Vermin, fungus and insects were major problems. Thatched cottages provide good thermal insulation, keeping the rooms cool from the sun and reducing the heat loss in the cooler periods. Heating of the dwelling and cooking would be via the open fire. Fuel even centuries ago would have been expensive. If you worked as a farmhand, miller, waterman, herdsman and so on you would not have had time or the transport to collect large numbers of logs for the fire. Buying them from a timber man or forester was the answer.

If you were born in the village, unless you worked you would not receive pay or keep the roof over your head. The landlord would only want you if you worked for him. If not, the dwelling would be needed for someone who could work. People did not travel far from their village and if they did a written document would be required by the squire as evidence that permission had been granted to travel. Freeman were the exception. The alternative would be as a trader or merchant with documents, a tinker or smithy who by trade would take their wares from village and towns. Finally a vagabond, a person of no fixed abode, no trade, no skill and with little to no possessions.

Where does all of this fit in with Acts and Regulations?

With a backdrop of peasant life working for the master, the lord of the manor, the squire, and the coming of the threshing machines, the steam engines and mechanisation in the fields it meant less employment. No employment equalled no home, no food or water – destitution. It is easy to see how the scouts could tempt the people to move to the glamorous new towns. How they would be disappointed. When they arrived the housing was dreadful. Terraces of back to back houses one up, one down became home to one or more families. The ground floor, often trodden earth, remained damp at best but was often a quagmire. Doors were taken off their hinges to lay on the wet ground, ladders gave access to the occupants of the room on the first floor. Sanitation was still a hole in the ground but was used by far more people. There was an open fireplace for cooking and heating if you could afford the fuel. No water pipes, just a communal pump close by. Food had to be bought from the mill or factory owner, often days old and rotten. Arid sulphur-filled smoke constantly flowed from the forest of chimneys. Like the landlords they had left in search of a new life, the new masters only paid if you worked, and if you couldn't pay the rent – eviction.

At the age of seven boys and girls worked in the mills and factories as cheap expendable labour. The dust, noise and dangerous working conditions and poor lighting, the open coal fire at the ends of the mills gave little heat unless you stood next to it. Girls with long hair or loose clothing could be caught in the machines, pulling them into the latticework of slung drive belts, mutilating or even killing them. Life and conditions must have been hell. The housing for the workers were generally sited to the East of the factories as the prevailing wind across Britain is from the South West. This meant that there was no escape from continual thick sulphurous smoke bellowing across their homes all day and much of the night. Many suffered with lung problems and simple illnesses often had fatal consequences. Life expectancy had dropped to about three decades if you were lucky in contrast with the longer lives in the countryside.

Management and the professional classes had far better housing, in the northwest of the factory area, out of the path of the continuous plume of smoke and stench. The factory and mill owners would live in the airy southwest, while those less wealthy but still rich in contrast with the workers resided to the southeast. All the major industrial cities were set out in a similar manner.

As the machinery became more efficient, larger buildings became popular. Certain occupations were more dangerous than others. The massive brick built multi-storey cotton mills with air filled with cotton dust were very susceptible to explosion. If a loadbearing wall blew out, the structure would collapse. However in the late 1700s fire proofing became a consideration. In 1792–93 Strutt's calico mill in Derby had cast iron columns with timber beams coated in lime plaster to protect against fire. The 2.70 m spanning brick shallow vaulted floors further improved the fire protection. A year later a new flax mill in Shrewsbury designed by Charles Bage became the world's first true iron-framed building incorporating the brick arched floors, cast iron columns and cast iron beams in place of the timber beams.

The new technology of ironwork framing enabled larger spans of floor areas without the need for large loadbearing walls. That meant larger glazed openings to let in more natural light. Lighting was very expensive. Tallow candles were both costly and easily extinguished therefore oil lamps would be used. Whale oil was becoming more expensive and it would be many decades before mineral oil would become available.

The living hell

Now we have a growing population in the cities, greater wealth for some whilst others suffered living hell. The rivers were becoming so thick with pollution that disease once more became rampant. Cholera came to Britain in the early 1830s. Rife in India and spread to Russia via the trade routes, it was thought that sailors had been carriers of the bacterium without themselves being affected. The Privy Council (no pun intended) knew of the devastation the British troops had met with disease and set-up a Central Board of Health in 1805 after concern about yellow fever. St Petersburg in Russia had a major cholera outbreak therefore the new Board of Health required all ships arriving in London to be moored up in Standgate Creek near Deptford under quarantine for 10 days. It also applied to ships from other British ports such as Sunderland as they had a cholera outbreak in 1831. The disease spread to Newcastle and then up into the Scottish ports with other outbreaks in Wales and up the west coast. The first recorded epidemic was in 1831–32 when 6536 died in London alone. The first outbreak took hold in the Limehouse area of East London. Sailors from the colliery ships from the North of England may have brought the disease. Sixteen years later cholera returned again in 1848–49 when a further 14 137 died, of which 10 738 lived in London.

Edwin Chadwick's Sanitary Report 1842 suggested that:

- corporate boroughs should take responsibility for drainage and water supplies
- non-corporate towns should set-up their own local Boards of Health
- taxation should be increased to pay for improvements.

The report became the basis for the Public Health Act 1848. However, as with many Victorian issues, people with money and power were reluctant to make any changes. The Municipal Corporations Act 1835 promoted about 250 towns to receive the Royal Charter and become councils or corporations. It was designed to enable the creation of Local Authorities, giving them powers to control the sewers and drainage in their areas. Many of them were very corrupt and the local businessmen held the official office. (For those who are interested it is worth reading copies of the Parliamentary Papers of 1835 for the Royal Commission on Municipal Corporations.) Most of the councils were Tory-led and in 1833 the Whig Government set-up a Royal Commission under the investigative leadership of a lawyer named Joseph Parker. He found that of the 285 towns investigated, most were unsatisfactory.

A previous 1832 Reform Act had seen the abolition of the 'rotten boroughs'. Corruption had been rife. Boroughs were self-governing towns that had the right to put forward a Member of Parliament. The commoners would in theory represent the views of the local people within the borough. However, to become elected most people had to have their own business to afford the time to become council members and/or an MP. Only men could vote and there were restrictions. The result was that mainly Tories took over the boroughs. As the Industrial Revolution took hold, previously wealthy boroughs became shadows of their former glory. Wealthy people who held office would pass their rights onto their sons to carry on their hold over the community. At one point 293 of the 405 elected MPs had been put forward by less than 500 voters each. The Reform Act 1832 became law in 1835 with help from Sir Robert Peel and the Duke of Wellington.

The Sanitary Act 1847 gave Local Authorities more powers to control the connection of sewers and drains to houses.

In 1848 the Public Health Act set-up a General Board of Health with powers to form local boards. The introduction of the regulation of building started. Both new and existing dwellings were to be provided with water and drainage and no new building could take place until the board had been informed where both the privies (WC) and the drains would be placed. The Act introduced a minimum height of habitable rooms or 8ft (2.44 m) in an effort to reduce stale air in the rooms. The requirement had a knock-on effect of adding more brickwork to all the walls, thus increasing building costs. Another updated Act regarding the built environment was the Nuisance Removal and Disease Prevention Act 1848. In the same year the Metropolitan Sewers Commission was established and gave permission for house drains and cesspools (open-topped cesspits) to be connected to the main sewers.

With more trade came more disease

Victorian Britain became a powerful global force, with almost 25% of the world's population under British rule. Trade with all parts of the globe meant greater wealth. It is not true that Britain only exploited countries (however we cannot go into that in this book). Trade did bring with it problems of disease. The third and largest cholera epidemic hit London in 1854. At the time the medical profession still continued to proclaim that disease was spread by bad air, termed 'miasmas'. Several notable medical speakers were questioning the establishment and had put forward their theories about the spread of disease. One such doctor, Dr John Snow, in September 1854 proved that cholera bacteria were spread by contaminated water. In 1854 a large cholera outbreak had killed 32,000 people across Britain in a matter of three months, 10,675 people in Gateshead and Newcastle alone. Snow sat in the local pub in Soho and noticed that people who drank beer were not affected, people who drank neat spirits were not affected, but people who had water with their drinks almost all died within days. That was the decisive point and Snow had the handle on the local pump at Broad Street removed and that outbreak ceased. The local leaking sewer had been about 7.00 m deep and the shallow well point only 9.00 m deep enabling the contamination to take place. It took several years before the medical profession accepted Snow's theory that cholera was a bacterium spread by contaminated water. He continued his research and traced another and perhaps the greatest source of contamination back to the privately owned Vauxhall Water Company. Water companies at that time removed the untreated water from the Thames and pumped it through a network of pipes and conduits to thousands of homes.

As one person became infected, his or her effluent contained the bacterium and entered the sewer systems. (According the World Health Organisation cholera only affects about 25% of those who carry the disease.) At that time it is estimated that the City of London had over 12,000 cesspits and in London town there were over 200,000 cesspits for a population of about 2 million people. The cesspits leaked, as did the inadequate foul water sewers. The recent re-invention of the flushing loo (WC) had meant that millions of litres per day were flushed into the cesspits, adding to the problem, most of which seeped or cascaded into the rivers and eventually the River Thames. The old London Bridge had been built on a series of small caissons, creating a dam effect and preventing the tidal currents from dispersing the effluent, waste from the Native Guano Company, dead animals and a soup of filth. In 1823 a less bulky new stone London Bridge was built, with the old one being removed in 1831, allowing the water to move more freely. Yet the width of the Thames still created drag to the currents, causing the effluent to beach and settle along the riverbanks. Cholera continued to be brought in, with another major outbreak in East London in 1866.

Human water carriers had been replaced by piped water pressurised by large steam-driven pumps. London was home to many wealthy people who in Victorian times wanted modern technology. The water closet became popular. It is a myth that Thomas Crapper invented the flushing loo. There is evidence that the Romans used water to sluice their latrines 2000 years ago. However, Victorian flushing loos had either taps that people left open or poorly designed water valves that continually leaked water, virtually emptying the reservoirs of north London. There was no metering of water therefore affluent Victorians wasted the precious water. The water, though pressurised, at best discharged as a trickle, therefore very inefficiently clearing the effluent. (Loo paper wasn't introduced until the late 1880s from America.)

In response to the massive waste of water the Metropolitan Water Act of 1870 required water systems to incorporate 'water-waste preventers'. Thomas Crapper was a trade-apprenticed working plumber who spent most of his time maintaining and fitting new water devices especially for sanitation. Popular history claims he invented a water closet that only allowed a single flush based on a siphonic system only discharging a specific amount of water. The discharge speed significantly increased, thus clearing the effluent more efficiently. However, according to Adam Hart-Davis it is very unlikely that Thomas Crapper invented the system. During the 1880s there were about 20 patents for siphonic systems per year, but no records of any patent for the siphonic flush by Thomas Crapper.

In London many of the rivers including River Fleet, Tyburn and Walbrook all flowed their stench-filled water into the River Thames. The Fleet is said to have been almost red with the blood of slaughtered animals, thick with fat from the soap factories and waste from the bone boilers and animal urine from the tanneries. For several decades the condition of the rivers had become worse, full of effluent, waste from the multitude of manufacturing works and rubbish from the docks and markets. Several eminent men complained to the authorities to no avail. Michael Faraday tried an experiment of dropping small pieces of white card into the waters at each pier to find that they were not visible below 25 mm from the surface.

The 'Great Stink'

The problem was becoming worse. After much deliberation and dispute with Sir Benjamin Hall the Metropolitan Board of Works invited entrepreneurs to put forward plans for a new sewage system for London. Bazalgette as Chief

Engineer in June 1856 put forward his plan to build two massive brick sewer tunnels both sides of the Thames and outfall down river where the tide would disperse the waste. It wasn't until 1858, the hottest summer on record, that even Parliament had enough of the stench. They had sheets of material soaked in deodorising chemicals derived from coal tar draped at all the windows on the river side. Eventually Parliament moved a few miles up river to escape the 'Great Stink'.

The Prime Minister of the time Disraeli put forward the Metropolis Management Amendment Act 1858 giving authorisation to Bazalgette to commence work. Not content with the massive sewers programme, Bazalgette put forward for a comprehensive programme of street improvements.

The 1848 Public Health Act required each district to establish a Central Board of Health if one tenth of the taxpaying population petitioned for it, or where the mortality rate exceeded the national crude death rate of 23 per 1000 over a period of seven years. Later in the same year an amendment to the 1848 Public Health Act ended the General Board of Health in favour of passing the duties back to the Privy Council and the Home Office.

The Health and Morals of Apprentices Act 1802 was the forerunner to the first Factories Act 1819, with revision in 1844 and 1847 limiting the hours of labour to 63 hours per week. This was further reduced to 58 hours per week in 1848. Child labour was commonplace: down the mines, in the mills, and the thousands of factories of the new Industrial Revolution. Children of parents would be put to work as young as seven being paid a mere pittance. If the parents gave permission the new laws of working hours could be ignored thus increasing the family income. Many children who survived the 50% mortality of the time and lived beyond 5 years old came from penny-less parents or orphans. They were classed as children of paupers therefore the laws did not protect them at all. They did not need paying and mill owner especially wanted the free labour to maximise their profits. This is one of the main reasons that many laws were very reluctantly passed in favour of the workers.

The 1867 Sewage Utilisation Act gave greater powers to the Local Authorities to dispose of public sewage. In the following year, the 1868 Sanitary Act increased the powers of the Local Authorities to dispose of sewage. The problems were ever-increasing.

1872 Public Health Act required every Local Authority to appoint a Medical Officer of Health. His job was to oversee the current Regulations and to ensure they were implemented.

The Public Health Act 1875 required urban authorities to make bye-laws for new streets, the stability and structure of new buildings and the prevention of the spread of fire. In the same year came the Artisans Dwelling Act 1875 (artisans are craftsman and tradesmen). The Act was to help the mass of town population who lived in run down and slum conditions. Compulsory purchase orders from the newly formed Local Authorities enabled whole areas of slum clearance and redeveloped as opposed to previously just individual buildings. In 1873 a document named the 'Return of Owners of Land' disclosed that less than 7000 men owned more than four fifths of the land.

Continuing with sewage, the Rivers Pollution Act of 1876 prohibited the disposal of solid matter, liquid or solid sewage directly in to rivers. The

Victorians then disposed of raw sewage on the land. Large lagoons stored sewage, separating the solid by settlement. The liquor was drained off and the solid matter dug out by hand and spread on the land. To sanitise the process the Victorians called the areas sewage farms, promoting a more healthy approach to disposal. They even had open days where interested people could visit the farms.

The Public Health (Water) Act 1877 was designed to simplify Local Government's purchase of private water works and required all rural sanitary authorities to ensure all dwellings to have wholesome/potable (drinking quality) water within a reasonable distance.

The Public Health (Amendment) Act 1889 related to more detailed reference to a number of sanitary and safety matters, in particular the new telegraph wires, the disposal of chemical waste, safety on building sites, and pleasure grounds/areas.

The Ribbon Act 1936 required consent by the Local Authority for any building within 220 ft (67 m) from the middle of any classified public road. Thirty-four county councils applied for the resolution restricting a further 13,000 miles of unclassified roads.

Why was the Town and Country Planning Act 1947 brought about?

The Town and Country Planning Act 1947 was brought in by the Labour Party directly after the Second World War. Prior to that date the control of development was mostly uncontrolled. Land owners could erect structures on their land as long as they complied with the Local Authority bye-laws for construction. After the war thousands of troupes returned to Britain to start their own homes however several years of bombing had meant there was a shortage of homes and materials. The Government required all Local Authorities to produce a building plan for their area and regulate what could be built on the land. There was a need for housing, food and work places. The Local Authorities had to allocate the land in their area to meet the needs of the local population and for the good of the country as a whole. Urban sprawl was joining towns by tentacles of development. The Labour Government where necessary compulsory purchased land and commissioned a programme of Council Houses on estates.

The birth of the Building Regulations

The first Building Regulations (1965) to become nationally accepted in England and Wales came into effect in February 1966 with the exception of Inner London Boroughs. London has since William I (1066) has maintained an independence from the rest of England. The Boroughs of the former LCC (London County Council) have London Building Acts that are in most cases more demanding than the Building Regulations.

The Building Regulations 1965 took over from the Public Health Act 1961 and many, but not all, of the building bye-laws of the various parishes, boroughs, district and county councils. Before the regulations came into effect it was possible for a designer to submit a set of plans into the Local Authority controlling one borough and have the plans approved. Across the road and under the control of another borough, the plans would be rejected. Architects and other professional designers who had commissions in areas they had not worked before found it very difficult to comply with all of the bye-laws as they were not generally written in one document. The advantages of the one package of rules to cover the whole of England and Wales was a very positive forward step. Writing any set of rules, or even a book, often makes the writer want to change the text after it has been published. Although everything is proofread several times, some things need re-writing. With the Building Regulations 1972 they were the actual documents of the Statutory Instrument. In other words the document had to go through the processes required by Parliament to make them law. This is in contrast with the Building Regulations today. The Building Act 1984 is the legal instrument. Quite a thin document by comparison it sets out the objectives of health and safety required in new, and alterations to existing, buildings.

When major incidents happen in new buildings they are thoroughly investigated and a Government report known as a 'White Paper' produced. In 1966 a new system build tower block was started in Newham, East London; Ronan Point. The 1960s had seen programmes of Local Authorities clearing post-war prefabricated houses and housing stock that was in very poor condition to make way for tower blocks of flats. Speed of erection was paramount as the residents of the homes to be demolished had to be temporarily housed, therefore 'system building' appeared to be the answer. The technique, used in Denmark, was based on precast reinforced concrete panels stacked on top of each other like a pack of cards. The panels were factory-made in steel moulds, ensuring accuracy. The edges were slotted for ease of connection and held in place by location bolts and cement grout. Taylor Woodrow Anglian had constructed about 2000 dwellings under licence and Larsen Nielson had built about 30 000 dwellings in Denmark over the previous 15 years, so the system was well proven. The 64 m high tower of 110 flats was completed in under two years and handed over in March 1968. On 16 May 1968 a gas explosion in flat 90 at 5:45 am blew part of the wall panel out on the south east corner with the consequence that the floor and wall above also collapsed. As the panels dropped a domino effect continued down the building, taking with it the whole corner from roof to ground level. Miraculously only four people died and 17 were injured, mainly due to the time of the gas explosion and the fact that the collapsed rooms on each floor were the sitting rooms. The cause has been a leaking gas cooker. The lady occupant had lit the gas to boil a kettle for an early morning cuppa.

From that disaster the 1972 Building Regulations Part D: Structural Stability, had a new requirement to stop any new building having a similar fate. The new regulation D20.(4):

> A building to which the provisions of this regulation apply shall be so constructed that if any portion of any one structural member (other than

a portion which satisfies the conditions specified in paragraph (5)) were to be removed—

(a) structural failure consequent on that removal would not occur within any storey other than the storey of which that portion forms part, the storey next above (if any) and the storey next below (if any); and
(b) any structural failure would be localised within each storey.

(Crown copyright)

As with all Building Regulations in modern times, they cannot be enforced retrospectively. The structural engineer's report suggested modification of every flat with additional metal strappings to prevent any future collapse, but public confidence had been lost and many people refused to live in the towers. Ronan Point was demolished in 1986. Although the design was blamed for the collapse, as far as I know there have not been any other disasters in the UK or Denmark.

In the 1960s and 1970s fast building methods also brought about two main issues:

1. Poor quality control on site – There had been a clerk of works on site who would have been inspecting the work as it progressed, so how did the bad workmanship take place?

   The construction workers had not provided the correct bearing surfaces for the loadbearing external walls. A minimum of 100 mm is required and on inspection after the collapse many of the joints were substantially less. An engineer at the time claimed that a 16 stone man could have barged the wall off its bearing as the joints were not correctly fitted.

   When Ronan Point was demolished the full extent of the bad practice was revealed. Joints that should have been grouted in were packed out with rubbish.
2. Technically poor designs – The building system had not been designed for structures over 6 storeys; the new towers were 20 storeys. Local Authorities were encouraged by national Government to build high. The 1956 Housing Act introduced subsidies to Local Councils for every floor built over five storeys.

   The Danish design had not been modified to meet the UK Building Regulations – how did it get Building Regulation approval?

A major re-write

The first major re-write of the regulations came with the Building Regulations 1972, comprising 188 pages of rules grouped in sections 'Part A to Part Q' (see Figure 7.1)

- Part A – Interpretation and General covered
- Part B – Materials
- Part C – Preparation of Site and Resistance to Moisture

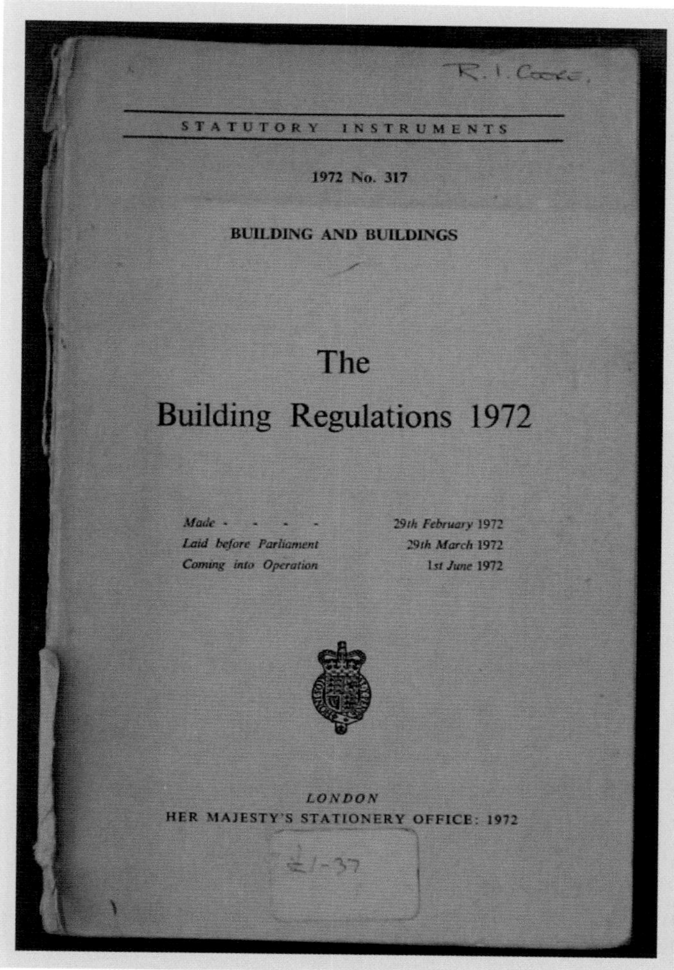
STATUTORY INSTRUMENTS

1972 No. 317

BUILDING AND BUILDINGS

The
Building Regulations 1972

Made - - - -	*29th February* 1972
Laid before Parliament	*29th March* 1972
Coming into Operation	*1st June* 1972

LONDON
HER MAJESTY'S STATIONERY OFFICE: 1972

Figure 7.1 Building Regulations 1972.

- Part D – Structural Stability
- Part E – Structural Fire Precautions
- Part F – Thermal Insulation
- Part G – Sound Insulation
- Part H – Stairs and Balustrades
- Part J – Refuse Disposal
- Part K Open Space, Ventilation and Height of Rooms
- Part L – Chimneys, Flue Pipes, Hearths and Fire Places
- Part M – Heat Producing appliances and Incinerators
- Part N – Drainage, Private Sewers and Cesspools
- Part P – Sanitary Conveniences
- Part Q – Ashpits, Wells, Tanks and Cisterns.

Part F – Thermal Insulation covered two pages with nearly half the page stating what the application and interpretation of the new requirements. Set out as a series of subsections F1 – F7 the introduction to what we now call U values appeared. F2 – 'Interpretation of Part F' was written in scientific terms and many building professionals at the time found it very difficult to understand. Heat loss before the 1972 Building Regulations generally had not been considered.

> F2.'surface heat transfer coefficient' means the rate of heat transfer in watts between each square metre of surface and the surrounding air when there is a difference in temperature of 1 degree Celsius between the surface and the surrounding air;
>
> 'surface resistance' means the reciprocal of the surface heat transfer coefficient; and
>
> 'thermal transmittance coefficient' means the rate of heat transfer in watts through 1 square metre of the structure when there is a difference in temperature of 1 degree Celsius between the air on the two sides of the structure
>
> (Crown copyright)

F3. Roofs now had to have a 'thermal transmittance coefficient of less than '1.42' in conjunction with any such ceiling'. Note no units of measurement were used. In F2 the word 'Watts' had a lower case letter although Watts is a person's name.

F5. External walls comprised five lines in totality, concluding 'the thermal transmittance coefficient of the wall is not more than 1.70'.

As you can see the Regulations were informative and to the point. Perhaps not that easy to understand, especially where professional people such as architects had been in practice for many years and had not been informed how to perform the necessary calculations.

As previously mentioned it is almost impossible to write any document that does not need changing. The Statutory Instrument status meant that any changes had to be passed by Parliament. It was a year before the 'First Amendment' to the Regulations was issued and came into operation 31st August 1973. A hefty 39 page document of additional regulation and amendments added to 26 pages of the Building Regulations 1972 pertaining to Structural Fire Precautions. Fire precautions were still the most important issue from the first 'curfew' of 1067.

A second amendment came into operation on 31 January 1975, again passing through the legalities required for Statutory Instruments. The Second Amendment was made in November 1974 introducing the term 'U value' for section F: Thermal Insulation. The F2 amendment gave interpretation to the changes and for the first time units of measurement: $W/m^2\,°C$ (see Figure 7.2).

> 'U value' means thermal transmittance coefficient, that is to say, the rate of heat transfer in watts through 1 m^2 of a structure when the combined radiant heat and air temperatures at each side of the structure differ by 1°C and is expressed in $W/m^2\,°C$;
>
> (Crown copyright)

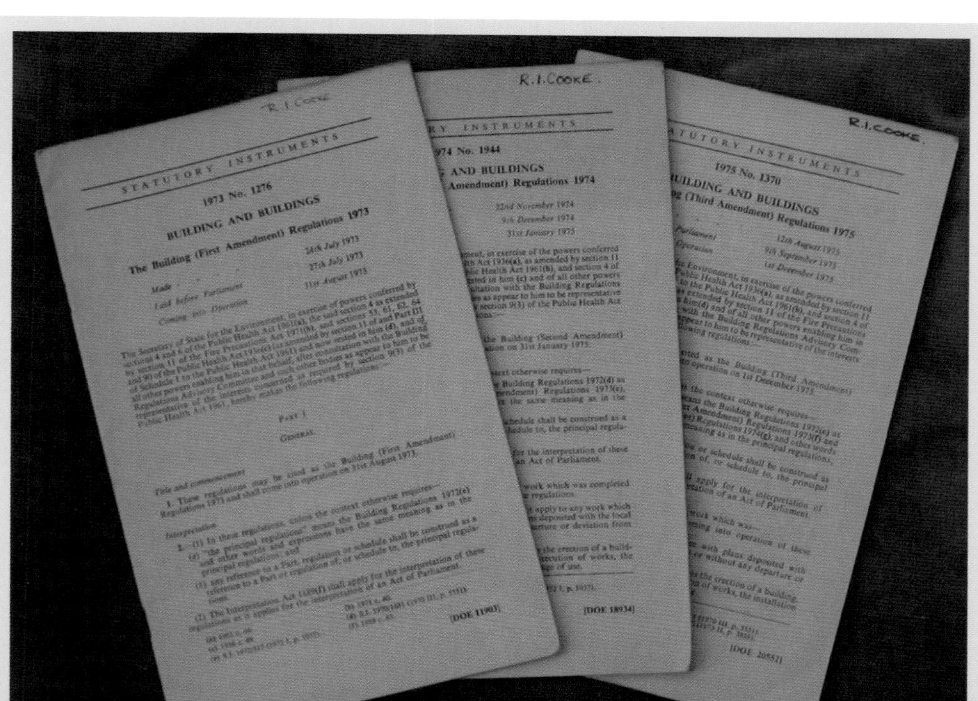

Figure 7.2 Building Regulation amendments.

The document presents the maximum U values for various elements of the buildings; walls, floors, roofs and perimeter walling. Note the absence of windows from the list. Windows were defined as:

> 'window openings' and specified as 'structural opening which is provided for a window irrespective of its size and function or for a hinged or sliding door or panel having a glazed area of 2 m^2 or more.'
>
> (Crown copyright)

and included in the 'wall boundary area'.

F3. then has a series of sub-sections:

> (2) The calculated average U value of perimeter walling (including opening therein) shall not exceed 1.8.
> (3) For the purposes of calculating the average u value of perimeter walling-
> (b) the U value of a window opening shall be assumed to be 5.7 if it has single glazing and 2.8 if it has double glazing; and
> (c) any other opening shall be assumed to have a U value equivalent to that of the wall in which it is situated.
>
> (Crown copyright)

Note there is no mention of cold bridging at returns or reveals.

To add to the confusion, the Table to Regulation F3 shows an elemental approach and item 1: External walls has a maximum U value of any part of element (in W/m^2 °C) of 1.0. Section F3 states not more than 1.8 for perimeter walling and the table on the following page also relating to F3 changes the wording to 'external wall' with a U value of 1.0. The remainder of the document refers to tables for roof structures in timber.

A Third Amendment came into operation a year later in December 1975 with minor changes to D: Structural Stability, and substantially more for E: Structural Fire Precautions. Part Q: Ashpits, Wells, Tanks and Cisterns under clause 32 was now to be omitted. That showed a change in approach to modern living. By then coal fires for the majority of dwellings had been replaced by cheap oil-fired or gas-fired central heating.

Clarification to terms used in the Town and Country Planning Act 1971 were made enabling terms such as 'greenhouse' to have a legal definition. Greater definition to the regulation of Part H: Stairs and Balustrades that became Part H: Stairways, Ramps, Balustrades and Vehicle Barriers. Further detail for:

> H3, (4) (a) subject to the provisions of paragraph (5), each tread (irrespective of whether its nosing is straight or curved on plan) is either a parallel tread or tapered tread; (b) subject to the provisions of paragraph (6), the rise of any step is uniform throughout its length and is the same as the rise of every other step in the flight;
>
> (Crown copyright)

The document then carries on for another seven subsections with additional sub-subsections.

Now that has all been cleared up, a series of 'Schedules' and 'Rules' for structural stability were included. As you can see the Building Regulations were evolving all of the time. A completely new set of Building Regulations came into force January 1977 (see Figure 7.3). The Building Regulations 1976, once again a Statutory Instrument, had become 307 pages of pure text containing more tables, schedules, and losing much, but not all, of the 'deemed to satisfy' and other legal wording that had punctuated its predecessor. Still there were no graphics, just black type on white paper. Some of the main sections had modified titles and Part Q had now gone. The list of twelve schedules had a similar word 'modification' but still remained mainly the same as the previous set of Building Regulations 1972. Part E: Structural Fire Precautions still had dominance in the document, emphasising the underlying issues that the regulations are primarily health and safety. Part F: Thermal Insulation; ostensibly remained the same with an inclusion that the designer could consider any lintel, jamb or sill associated with an opening in a wall as either part of the wall or part of the window. The thermal implications of 'cold bridging' had now appeared although not in name. Part G: Sound Insulation remained word for word as it appeared in the 1972 document. Part H: Stairs, ramps, balustrades and vehicle barriers now had clear definitions of specific words and terms. The inclusion of the

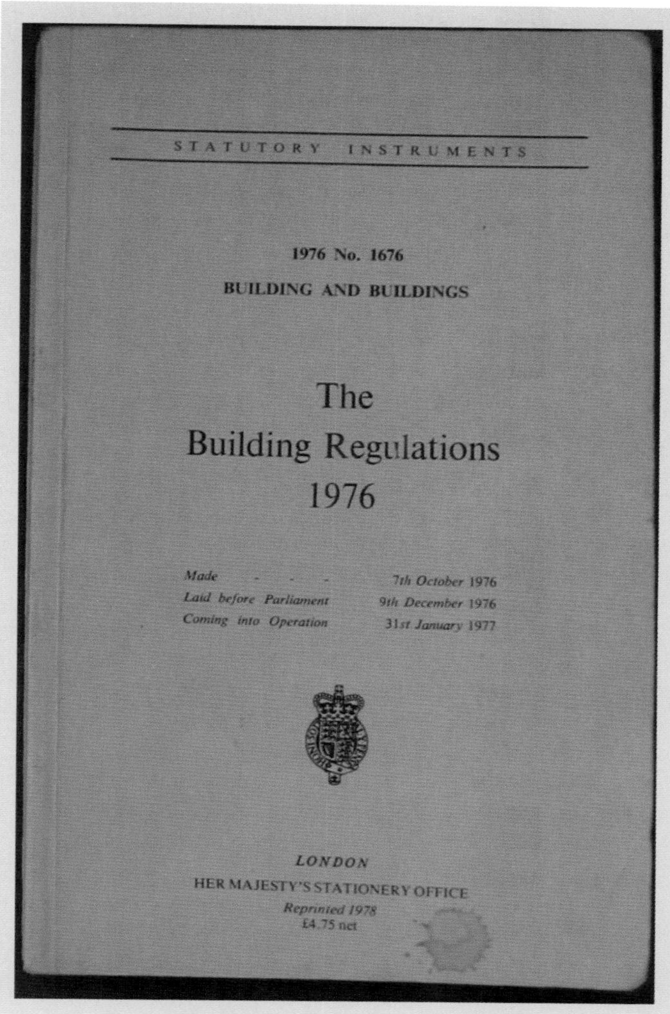

STATUTORY INSTRUMENTS

1976 No. 1676

BUILDING AND BUILDINGS

The Building Regulations 1976

Made - - - 7th October 1976
Laid before Parliament 9th December 1976
Coming into Operation 31st January 1977

LONDON
HER MAJESTY'S STATIONERY OFFICE
Reprinted 1978
£4.75 net

Figure 7.3 Building Regulations 1976.

provision of artificial lighting for stairways and landings and the means for operation became part of the Regulation. Tapered stairs known as 'kites' and 'winders' had rules for measurement. The 'going' of the tread was to be measured at the centre, ensuring that it followed the rules straight treads on a consecutive flight of stairs. This meant that 'winders' would no longer be possible. Winders are stair treads that are wide one end and reduce to virtually a few millimetres at the other. Very common on flights of stairs in old buildings, especially where space was at a premium, the stairs would be almost as a spiral and hardly any surface to step onto – Figure 7.4 and Figure 7.5.

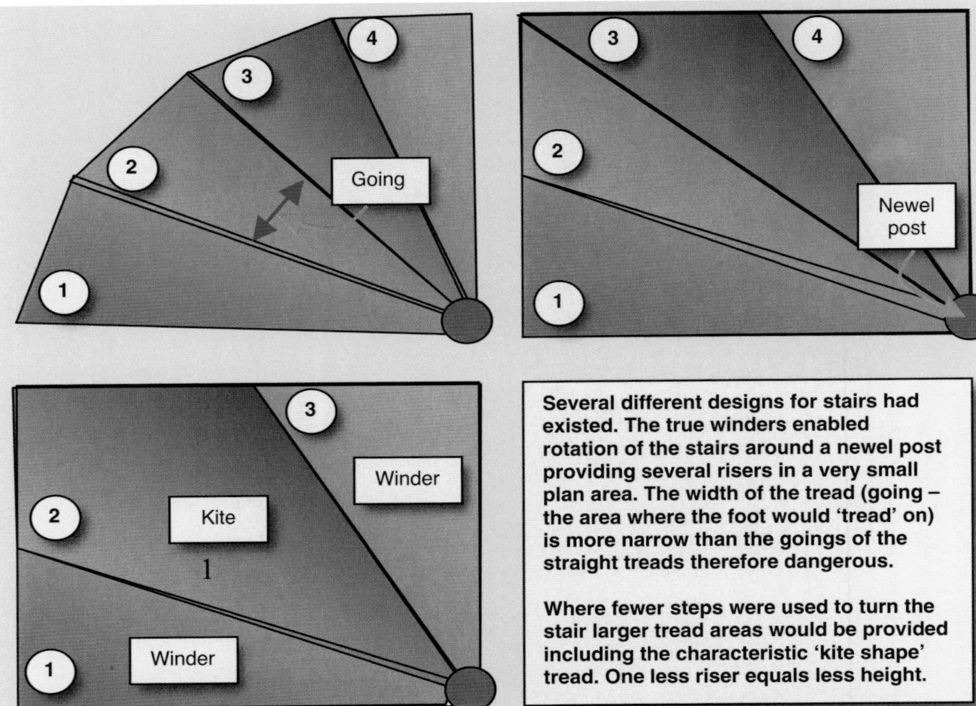

4
3
2
1
Going
3
4
2
1
Newel post
3
2
1
Kite
Winder
Winder
1
Several different designs for stairs had existed. The true winders enabled rotation of the stairs around a newel post providing several risers in a very small plan area. The width of the tread (going – the area where the foot would 'tread' on) is more narrow than the goings of the straight treads therefore dangerous.
Where fewer steps were used to turn the stair larger tread areas would be provided including the characteristic 'kite shape' tread. One less riser equals less height.

Figure 7.4 Stair 'winders' and 'kites'.

Figure 7.5 Stair 'winders'.

The price of oil

As with the 1972 Building Regulations, amendments followed. The most significant was the introduction of Part FF in June 1979 (see Figure 7.6). The world had seen an increase in oil prices from several of the major oil producing countries in the early 1970s. Cheap oil was a thing of the past unless you lived in America. Most oil central heating systems for domestic property were changing to gas-fired or electric. Those who kept their oil-fired boilers noted a large increase in costs. Many new housing estates had contracts with oil suppliers that prevented them converting to other fuels. British Gas and the electricity companies had run similar schemes that they would pay for the service mains to be run into any new estate if they had the monopoly. They were known as Gold Medallion Estates.

During the 1970s the major oil-producing countries of the world clubbed together to form OPEC (Organisation of Petroleum Exporting Countries). Between them they controlled 70% of all the oil sold. In 1973 they increased the cost of oil per barrel, which had devastating effects worldwide. (Note the date, and the date of the first Building Regulation amendment – August 1973.) Denmark, for example, had been almost 100% oil fire burning. Here in the UK many of the power stations had been converted to burning the once cheap oil and were now being converted to running on natural gas during the period known as the 'dash for gas'. In 1979 the Shah of Iran further increased the cost

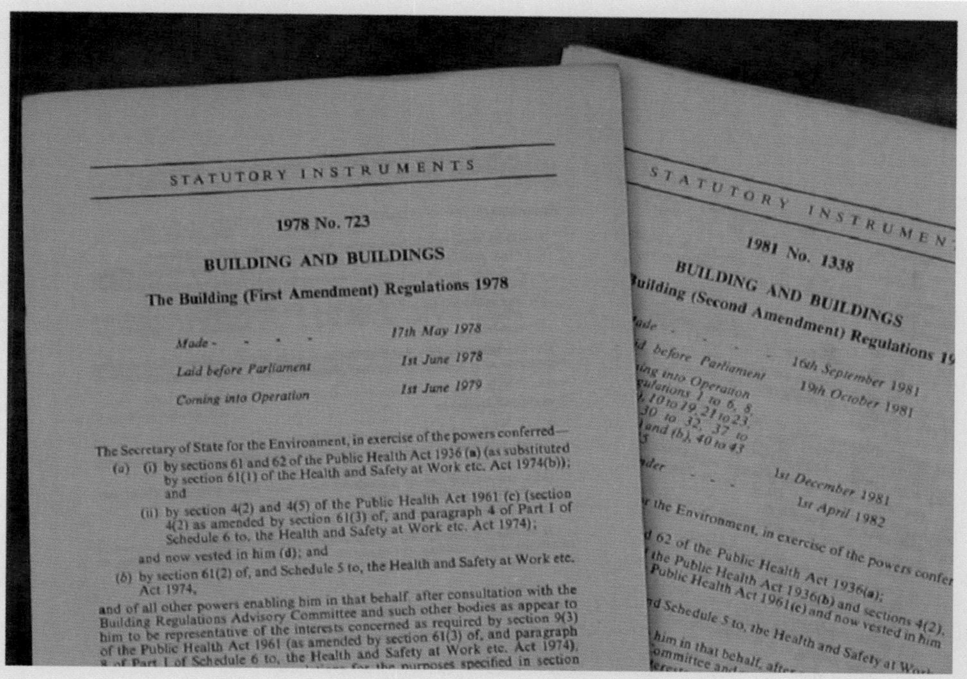

STATUTORY INSTRUMENTS

1978 No. 723

BUILDING AND BUILDINGS

The Building (First Amendment) Regulations 1978

Made - - - -	*17th May 1978*
Laid before Parliament	*1st June 1978*
Coming into Operation	*1st June 1979*

The Secretary of State for the Environment, in exercise of the powers conferred—

(*a*) (i) by sections 61 and 62 of the Public Health Act 1936 (a) (as substituted by section 61(1) of the Health and Safety at Work etc. Act 1974(b)); and

(ii) by section 4(2) and 4(5) of the Public Health Act 1961 (c) (section 4(2) as amended by section 61(3) of, and paragraph 4 of Part I of Schedule 6 to, the Health and Safety at Work etc. Act 1974);

and now vested in him (d); and

(*b*) by section 61(2) of, and Schedule 5 to, the Health and Safety at Work etc. Act 1974,

and of all other powers enabling him in that behalf after consultation with the Building Regulations Advisory Committee and such other bodies as appear to him to be representative of the interests concerned as required by section 9(3) of the Public Health Act 1961 (as amended by section 61(3) of, and paragraph 8 of Part I of Schedule 6 to, the Health and Safety at Work etc. Act 1974), ...

STATUTORY INSTRUMEN...

1981 No. 1338

BUILDING AND BUILDINGS

...uilding (Second Amendment) Regulations 19...

Figure 7.6 Building Regulation amendments.

per barrel. World politics were about to change. The once poor nations of the world were now flexing their oil-based muscle.

The Building Regulations quickly reflected the concerns, requiring U values to lower. Part FF: Conservation of fuel and power in buildings other than dwellings came into effect. Triple glazing at 2.0 $W/m^2\,°C$ was an added option under FF4. More detail about how to achieve better thermal insulation in a series of tables and schedules was given. The U values for external walls, floor with exposed under surfaces and the roofs now had to be better than 0.6 $W/m^2\,°C$ for buildings generally and 0.7 $W/m^2\,°C$ for storage areas. Trade-off could be used. For example, an external wall with windows that were single glazed would achieve a specific total loss of 'x', but larger windows could be used if double or triple glazed, so long as the total loss did not exceed that of the wall with single glazed windows.

In 1981 the Second Amendment to the 1976 Building Regulations took effect in two stages; December 1981 and April 1982. The U values for walls remained at 0.6 $W/m^2\,°C$ but heat loss through the roof was now capped at 0.35 $W/m^2\,°C$, almost half that of the previous requirement. Roof voids now required a lot more and more efficient insulation. There were two significant inserts, schedules 1 and 2.

1 Schedule 1 Part Q: Control of space and water heating systems
2 Schedule 2 Part R: Thermal insulation of pipes, ducts and storage vessels.

Part Q required control of intermittent heating (Q4) and control of the operational selection of boilers (Q5).

Part R required pipes, ducts and storage vessels to be thermally insulated. Pipes and hot water cylinders were generally bare and lost considerable heat therefore used more fuel.

Schedule 11 provided several systems of construction for external walls with the resistance (R values) and U values for ease of use.

Computers in industry

Bear in mind that computers were not in general use and extremely costly. The company I was working for had a whole corridor of computers with massive reels of magnetic tapes rotating back and forth. Technicians in white coats with dust covers over their shoes, airlocks and strict air conditioning pampered the banks of computers. In comparison with today, a small palmtop portable computer probably has more calculating power. To carry out sophisticated U value calculations involving solar gain and light hours etc. for commercial buildings the data had to be transmitted to Canada using ticker-tape feed and a telephone line. Software was still in its infancy. Handheld calculators were very basic therefore calculations were very time-consuming.

The Building Act 1984 is the law relating to buildings in England and Wales. It does not cover Scotland or Ireland. Part of the Act enables a set of rules known as the Building Regulations. They are mainly to do with health and safety. Unlike their predecessors the new Building Regulations are not Statutory Instruments. The Building Act 1984 is, though. The Act enables the

'Power to make building regulations' and 'continuing requirements' to be made, therefore any changes or amendments do not have to go through Parliament. The Building Regulations are only part of the Act. It is well worth reading the document online to see the actual legislation and quash many of the 'urban myths' relating to construction: www.opsi.gov.uk/RevisedStatutes/Acts/ukpga/1984/cukpga_19840055_en_1.

Take 'section 36' for example. Where construction work has been carried out either without permission or not to the requirements of the Regulations, the Local Authority can issue a 'section 36 notice'. The full extent of the law is shown in easy to read detail and is very useful information for students who want to become building surveyors or who are studying building law.

Today in Britain all new buildings must comply with the Building Regulations 2000 and amendments (there are exceptions such as Local Acts, but some of them have been repealed under the Building (Repeal of Provisions of Local Acts) Regulations 2003). In order to explain the regulations more fully, a series of 'approved documents' have been produced. Thin manuals, very well illustrated, these provide solutions that if followed correctly 'should' meet the Building Regulations. The approved documents are not 'law'. They are suggested solutions to meet the requirements of the Building Regulations. If you have another method or technique that will provide a solution, it must comply with the relevant British Standards, BS Codes of Practice or Euro Codes. The Building Control Officer or approved inspector may ask for evidence that your solution will fully comply, such as certificates from the British Board of Agrément (BBA) (see Figure 7.7), or other testing stations such as Building Investigation and Tests Services (BITS). It is very important to check with the issuing establishment that the certificates are still current, as companies advertising on the internet are still showing test certification from a test station that went into bankruptcy in 2004. Also, as products are modified they may no longer comply with the test certificate. The BBA actually recommend that you check companies who advertise they have a BBA certificate online at www.bbacerts.co.uk and use the 'search product section'.

At the time of writing the following list of Approved Documents are available. To ensure the latest versions are being used it is useful to check or read them free online using the following website: www.planningportal.gov.uk/england/professionals/en/1115314110382.html

- (AD) A – Structure
- (AD) B – Fire Safety
- (AD) C – Site preparation and resistance to contaminates and moisture
- (AD) D – Toxic substances
- (AD) E – Resistance to the passage of sound
- (AD) F – Ventilation
- (AD) G – Hygiene
- (AD) H – Drainage and waste disposal
- (AD) J – Combustion appliances and fuel storage systems
- (AD) K – Protection from falling collision and impact
- (AD) L1a – Conservation of fuel and power (New dwellings)

CI/SfB

(13.9) n6

Designated by Government to issue European Technical Approvals

Visqueen Building Products
Maerdy Industrial Estate
Rhymney
Tredegar
Gwent NP22 5PY
Tel: 01685 840672 Fax: 01685 842580

Agrément Certificate No 94/3009
*Third issue**

VISQUEEN BUILDING PRODUCTS DAMP-PROOF MEMBRANE
Membrane étanche à l'humidité
Feuchtigkeitssperre

Product

- *THIS CERTIFICATE REPLACES CERTIFICATE No 87/1846 AND RELATES TO VISQUEEN BUILDING PRODUCTS DAMP-PROOF MEMBRANE, A LOW-DENSITY POLYETHYLENE MEMBRANE FOR USE IN SOLID CONCRETE GROUND FLOORS NOT SUBJECT TO HYDROSTATIC PRESSURE, TO PROTECT BULIDINGS AGAINST MOISTURE FROM THE GROUND.*
- *The product is available in thicknesses of 250 μm, 300 μm and 500 μm.*
- *It is essential that the product is laid in accordance with the recommendations of clause 11 of CP 102 : 1973 or with this Certificate.*

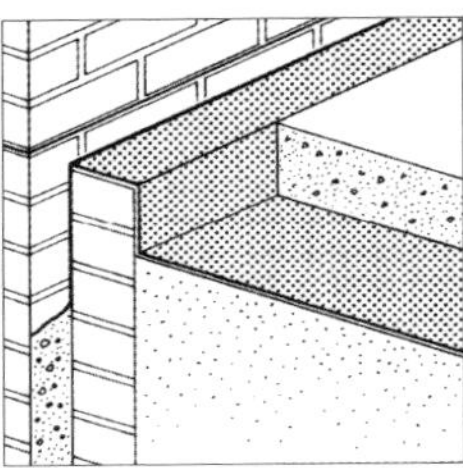

Regulations

1 The Building Regulations 2000 (as amended) (England and Wales)

The Secretary of State has agreed with the British Board of Agrément the aspects of performance to be used by the BBA in assessing the compliance of damp-proof membranes with the Building Regulations. Inthe opinion of the BBA, Visqueen Building Products Damp-proof Membrane, if used in accordance with the provisions of this Certificate, will meet or contribute to meeting the relevant requirements.

Requirement: **C4** — Resistance to weather and ground moisture
Comment: The product will meet this Requirement. See sections 8.1 and 8.2 of this Certificate.

Requirement: **Regulation 7** — Materials and workmanship
Comment: The product is an acceptable material. See section 13.1 of this Certificate.

2 The Building Standards (Scotland) Regulations 1990 (as amended)

In the opinion of the BBA, Visqueen Building Products Damp-proof Membrane, if used in accordance with the provisions of this Certificate, will satisfy or contribute to satisfying the various Regulations and related Technical Standards as listed below.

Regulation: 10 — Fitness of materials
Standards: B2.1 and B2.2 — Selection and use of materials, fittings, and components, and workmanship
Comment: The product complies with these Standards. See section 13.1 of this Certificate.

Regulation: 17 — Resistance to moisture
Standard: G2.6 — Preparation of a site and resistance to moisture from ground —Resistance to moisture from the ground
Comment: The product can enable a floor to satisfy the requirements of this Standard. See sections 8.1 and 8.2 of this Certificate.

3 The Building Regulations (Northern Ireland) 2000

In the opinion of the BBA, Visqueen Building Products Damp-proof Membrane, if used in accordance with the provisions of this Certificate, will satisfy or contribute to satisfying the various Building Regulations as listed below.

Regulation: **B2** — Fitness of materials and workmanship
Comment: The product is an acceptable material. See section 13.1 of this Certificate.

Regulation: **C4** — Resistance to ground moisture and weather
Comment: The product can enable a floor to satisfy the requirements of this Regulation. See sections 8.1 and 8.2 of this Certificate.

4 Construction (Design and Management) Regulations 1994 (as amended) Construction (Design and Management) Regulations (Northern Ireland) 1995 (as amended)

Information in this Certificate may assist the client, planning supervisor, designer and contractors to address their obligations under these Regulations.

See section: *5 Description* (5.1).

Readers are advised to check the validity of this Certificate by either referring to the BBA's website (www.bbacerts.co.uk) or contacting the BBA direct (Telephone Hotline 01923 665400).

Figure 7.7 British Board of Agrément Certificate.

- (AD) L1b – Conservation of fuel and power (Existing dwellings)
- (AD) L2a – Conservation of fuel and power (New buildings other than dwellings)
- (AD) L2b – Conservation of fuel and power (Existing buildings other than dwellings)
- (AD) M – Access to and Use of Buildings
- (AD) N – Glazing
- (AD) P – Electrical safety – Dwellings
- Regulation 7.

Town and Country Planning Act revisions

The Town and Country Planning Act has been updated several times since 1947, and is currently the Town and Country Planning Act 1990. Like the Building Act 1984, the TCP Act 1990 is a Statutory Instrument, i.e. law. The full document can be accessed online at www.opsi.gov.uk/ACTS/acts1990/ukpga_19900008_en_1 and makes interesting reading. For example, subsection '102: Orders requiring discontinuance of use or alteration or removal of buildings or works' shows that the Local Authority has powers to regulate the use of land and where necessary have buildings or works removed. This part of the Act enables Local Authorities to take legal action against those who illegally build or use buildings or land for improper use.

Subsection '172: Power to issue enforcement notice' states that the Local Authority has the power to challenge any breach of planning control after the end of 1963. For example a garage being used as a habitable room without the planning consent for change of use can be halted.

The LA can, however, insist that retrospective planning applications are made. If in their opinion permission would have been granted, for a set fee approval will be given. However, the work also has to comply with the Building Regulations and that would be a separate application. In cases where blatant disregard for planning or building regulations has taken place such as building on green belt land, the LA can take legal action through the courts to enforce the land being returned to its original condition at the expense of the builder and or owner.

Subsection 198: Tree preservation is part of the TCP Act 1990. Neglect of the maintenance of land is covered by subsection 215: Power to require proper maintenance of land. The regulation of advertisements is covered under subsection 220, whilst the compulsory acquisition of land for redevelopment and other planning purposes is subsection 226.

Part of the TCP Act 1990 contains Regional Spatial Strategies (RSS). They are legally binding documents that were designed to replace Regional Planning Guidance and Structure Plans. RSSs are formulated by the Local Authority but should be showing more regard to national interests. The Deputy Prime Minister as he was then, John Prescott, led the policy changes as he felt that the south east of England housing was not dense enough, and that Local Authorities were not making enough land available for housing. The outcome of the policy changes has seen much resistance from the local residents of South

Cambridge District Council and Cambridge County Council. The councils rejected plans for 6700 homes, one of the Government's 'eco-towns', to be sited on green field land. The development was to be on the Hanley Grange site and backed by the supermarket Tesco. The new planning policy allowed the planning process to be bypassed but the local residents were very vocal in their objections and Tesco withdrew its financial support. The development has now got to go through the conventional planning process and meet the Regional Spatial Strategy. At the time of writing no decision has been made. The same developer had intended to build 10 000 homes in Norfolk but the local residents once again strongly objected and at the time of writing (Sept 2008) it is considering pulling out.

To view the RSSs online you will have to register, as the documents are continually being updated. By registering you will be assured of the current RSS documentation.

In addition to the Building Act 1984 and the Town and Country Planning Act 1990 there are several water-related acts such as the Water Industry Act 1991 and the Water Acts 2003. The following websites enable the full documents to be inspected:

- The Water Industry Act 1991: www.opsi.gov.uk/acts/acts1991/Ukpga_19910056_en_1.htm
- The Water Act 2003: www.opsi.gov.uk/acts/acts2003/ukpga_20030037_en_1
- For general water-related Acts and Regulations the NetRegs website provides useful links: www.netregs.gov.uk/netregs/legislation/current/63616.aspx.

Chapter 8

Speculative housing

So far we have looked at the various stages of construction from the client having an idea, 'inception', through to the issues surrounding planning and production. In this chapter we will look at the scenario of a small domestic development in the context of the various stages. (Unfortunately the real project has been shelved due to the current financial climate therefore parts of real projects have been modified to create the likely course of events).

The issues of influence of the Town and Country Planning Act 1990 and various associated legislation together with covenants and 'rights of way' show that building is not just designing and production. Updates and policy changes by Governments and Local Authorities have been included, showing what is required and how to apply for planning permission and to show compliance with the Building Act 1984.

Testing and certification procedures have been included plus commentary and observations of the organisations responsible for monitoring the work. Concluding with an insight into the way building methods and materials have changed and the consequences, and finally the new house owner's 'building manual'.

The term 'speculative' is used where the client intends to proceed with a development without having a specific final buyer. For example, a speculative office block would be constructed and then placed on the market for sale. Likewise a speculative housing estate would be developed and each building placed on the market, usually for sale. If the developer intends to lease or rent the property and not sell it on, then technically it would not be speculative housing as the end buyer, the property owner, would be known – the developer would be the owner.

Land for development traditionally has been the least technically or socially challenging. However, today, as land becomes ever more difficult to buy, all land and property is considered. A developer's objective is to maximise the profit on the smallest outlay. If a plot of land has not previously been built upon, or has not in living memory been built upon, it is termed a 'green field site'. The ground/soil will not have been compacted and natural stratum will be undisturbed. (The reason for the clarification about living memory refers to land that had been previously used, say, prior to the Second World War and the buildings destroyed, rubble cleared and the ground reinstated to park or woodland. The land may appear to be a green field site yet has foundations and services buried beneath the surface.)

The speculative housing project used as this example has been built on a brown field site. Brown field sites are plots that have obviously been previously developed. For example, in the past four decades British manufacturing has changed from large labour-intensive factories built in the previous century to more modern purpose-built structures. In London the Metal Box Company had several very large factory units. They are no longer needed therefore developers bought the old factory units and converted them to smaller self-contained factory units termed 'nursery units'. The land was a 'brown field' site.

Developers as previously mentioned need to maximise the profit on the development therefore carry out extensive pre-contract research. If the land has been previously used for industrial or commercial purposes there may be contamination in the soil. Contaminated soil may have to be removed, capped or encapsulated, all of which can be very expensive. A good example is the Dome in Greenwich London. Built on the site of a very large coal gas works, the soil was so heavily contaminated with chemicals that it was not commercially viable to develop. The cost of removing the contaminated soil that included arsenic, cyanide, asbestos and a mixture of other hazardous materials would prohibit making a profit on developing the site. The Government used taxpayer's money to pay for cleaning the site and removing the contamination prior to building the Dome. Although it was a temporary structure, future development of the site will now be possible and commercially viable.

8.1 Rayleigh Road project

The proposed site comprised eight two-storey terraced flats built in the late 1930s. Where flats share the same plot of land it is normal practice for the landlord to remain owning the land and lease it for development. The developer may as in this case build purpose designed two-storey flats – more correctly termed maisonettes – for sale leasehold. The person who buys the property will then hold possession of their part of the structure until an agreed date where the landlord can then decide what he or she wants to do with the land. Leases can be 99 years from the original development therefore as each leaseholder sells their interest the value decreases. However as property prices over the past four decades have increased substantially, the value of the lease appears to increase. When the lease nears its end then only a short-term interest can be achieved and the value is reflected in the selling price. The value is only based on their share of the interest. It does not include the land or any other part of the structure such as the roof unless it is the top storey.

Behind the terrace of maisonettes there were two old wooden barns, remnants of the farm that had previously covered the land. The barns had been used as commercial workshops and storage areas. The tenant had died due to a road accident and the building was left derelict and to decay. The land had not been sold since 1972 therefore there had been no compulsion to register the land under the Land Charges Act 1972. Under the Land Registration Act 2002 there is a legal requirement for unregistered land termed 'legal estate' to be registered under the Housing Act 1985. Also under the Housing Act Part 5 (the right to buy) a tenant could negotiate with the landlord to buy the land that the structure stood on. The transaction is complex and should be left to interpretation by legal professionals.

The developer had bought several of the maisonettes from the tenants over a period as they became available on the open market and then rented them out until acquisition of all of the property could be achieved. The owner of the old barns and the land they stood on was traced via a search on the Land Registry for a set fee. The next stage was to establish whether there were any covenants, rights of ways or easements on or over any of the property. A covenant can be placed over land as to its future use. In this example the landowner farmer had decided to sell part of his land to a builder developer for housing long before the Town and Country Planning Act existed in the 1930s. As part of the contract the farmer could have written in that the builder may only build seven bungalows on the land and that any subsequent home owner cannot store or park a caravan or motor home within the curtilage of the original plot. This would be termed a 'Restrictive Covenant' and can be enforced by the farmer or his dependants or in the case of a company the covenant can be held and sold.

Restrictive covenants can go back over a century therefore it is essential that they are found during the searches prior to development. However, enforcing the covenant may not be a straightforward procedure. For example, my own house was built on farmland in 1955. The covenant states that the dwelling must exceed a specific value to prevent a shack being erected and that the structure be a minimum of 30 ft (about 10 m) from the east boundary. At the time there was no requirement for a road to be constructed although space had to be allocated for a road in the future. (Today property must have a road prior to development although the road may not be adopted by the Local Authority and would be designated as a private road.) Other restrictions include that

I may not erect a bone factory or allow a travelling circus or caravan to enter the plot. Who can actually enforce the covenant now is not clear as the originator must be long dead and unless the dependant family know about and wish to invoke such restrictions they are unlikely to be effective. They still exist though.

Rights of way have logical origins, some dating back centuries. Imagine a large area of land such as a farm. You buy a plot of land to the east of the farm and want to visit your friend who lives to the west of the farm. To walk from your house around the farm on the public highway may take 3 hours however if you could walk through the farmer's fields it may only take 20 minutes. The farm land is private and to access and pass over the field is illegal without the permission of the farmer or his/her agent. You may have seen signs that state 'trespassers will be prosecuted'. Normally the trespass part of entry unless occurring regularly would not be cause for redress. However, if, say, the action damaged a crop by walking through a cornfield or if walking a dog that then worried live stock the farmer rightfully can have you legally removed and if necessary prosecuted for the damage by trespass. In contrast if you get to know the farmer and he/she says that you can walk through the farmyard and along a specific pathway through the field there is no problem. However, if the permission is used for a long period and perhaps your children continue the action, a 'right of way' may be established over a long period of time. More commonly rights of way are written into legal contracts that form part of the deed of covenant.

Rights of way can be classed as 'easements'. They can be implied, express or prescriptive. At the end of my garden a 12 foot (4 m) wide strip of my plot cannot be planted or obstructed as my neighbours have an easement or private right of way across my garden. It is a prescriptive right that states they can pass and re-pass with or without horses and/or a cart at any time, 365 days of the year. The land is still part of my plot therefore private land. I am responsible if negligently I leave something on the strip of land that causes an accident or harm to my neighbour whilst passing over my land. If, however, my neighbour should stop or take use of my land, he or she would breaking the right of way easement. For example they could not plant a hedge or erect a gate or fence on my land. It is out of the scope to go further into the legal aspects of rights of way, and easements although the subject can be very interesting.

So far we have:

- A developer who has over a period of time bought property and acquired title deed over adjacent land with and intention of building a small housing estate.
- The development is all brown field sites and it is known that no industrial buildings or landfill has taken place.
- The developer has employed a solicitor to carry out a full search over the land to establish whether any easements, rights of way or covenants are in existence.
- The developer had consulted a local Town Planning firm for advice as to whether the proposed speculative housing project would be possible

under the Town and Country Planning (Local Development) (England) Regulation 2004.* As the number of dwellings per hectare did not meet the 30 dph (dwellings per hectare) minimum requirement, an outline planning application was made using the form 'Application for Outline Planning Permission With Some Matters Reserved. Town and Country Panning Act 1990'. (Blank forms can be downloaded from any Local Authority website.)

Previous legislation and policy were cancelled on 1 April 2007, including:

- the Town and Country Planning (Residential Development on Greeen-field) (England) Direction 2000
- the Town and Country Planning (Residential Density) (London, South East England, South West England, East of England and Northamptonshire) Direction 2005.
- General Policies and Principles Planning Policy Guidance Note 1 (PPG1) February 1997
- Planning Policy Guidance 3: Housing (PPG3) March 2000.

All of the above are covered in the:

- Planning Policy Statement 3 (PPS3): Housing of 29 November 2006 (with effect from 1st April 2007)

An assessment of how the PPS3 will affect different social groups has been set out in PPS3: Housing – Equality Impact Assessment (9 May 2007). The 9 page document highlights the needs for dwellings of disabled people, families with children, single parent families, homeless households, older people, and students. Dwelling densities have an impact on various groups in society. For example, high levels of density may be considered a problem where access is required for people with mobility issues such as the elderly or physically disabled. However, similar issues exist for families with very young children, therefore when planning applications are made for dwellings the list above would be taken into consideration as part of the Local Authority Planning Policy Development Plan known as the Local Development Framework (LDF). A useful guide to building higher density developments 'Better Places to Live by Design: A Companion Guide to PPG3' can be bought from Her Majesty's Stationery Offices or downloaded free:

www.communities.gov.uk/publications/planningandbuilding/creating-betterplaces

- Planning Policy Statement 1 (PPS1): Delivering Sustainable Development
- Supplement to PPS1: Planning and Climate Change (17 December 2007)
- Planning Policy Statement 12 (PPS12): Local Spatial Planning (4 June 2008)
- Building a Greener Future Policy Statement (July 2007)
- PPG2 Green Belts (May 2006).

* The full document can be viewed at www.opsi.gov.uk/si/si2004/20042204.htm.

The full documents can be downloaded free from www.communities.gov.uk/documents/planningandbuilding/pdf/planningpolicystatement3.pdf and www.communities.gov.uk/publications/planningandbuilding/ppsclimatechange. There are links to the other documents from either of these websites. The most commonly used documents have been listed above, but there are many more specialised documents available.

The Town Planner acting as agent to the developer made an informal meeting with the Local Authority Planning department where general opinion of the proposed scheme could be discussed. The comments given by the Local Authority Planner are not legally binding and cannot be taken as the opinion of the Local Authority; they are more of a guide as to the possibility of the scheme. If the proposal has no chance of obtaining planning permission then in most circumstances that will indicate that no further work should be carried out. There are exceptions though. For example, a project may be granted permission if it includes other work that will benefit the Local Community such as a new road or traffic calming scheme at no cost to the Authority.

In this case the number of new dwellings was debatable therefore an Outline Planning Application was submitted. The application comprised:

- Floor plans and elevations of the proposed building work to a scale of 1:50 for detail, and 1:100 for elevations and plan layouts. Colour and fabric (the proposed building materials such as 'red facing brick', 'brown interlocking tiles' etc.) noted on the drawings. At this stage there is no need for specific detail. The planners need to know the overall concept of the proposed development. Where protected trees or trees in adjacent property exist they should be shown on the drawings including species and approximate height and spread of the trees. All boundary definitions should be shown on the elevations such as walls, post and rail fences and gates especially if tall and adjacent to the public highway.
- Cross sections where applicable of both the site and the buildings. This is especially important where the site is on a gradient.
- The site plan, also known as the 'block plan', should be to a scale not less than 1:500 and include roads and existing trees of all adjacent property. Commonly a smaller scale of 1:1250 or 1:2500 is used showing the project plot curtilage lined in red crayon and the area shaded in red, and adjacent property shaded with a blue crayon – see Figure 8.1. If there had been any easements or public rights of way over the plot they would have been outlined in green crayon. Application forms are available from the Local Authority controlling the area or in most cases applications can be carried out electronically using the e-planning applications via the Planning Portal website. Enter the postcode for the proposed development and the website will directly link you to the appropriate Authority.
- Completed 'standard planning application forms P1' has now been replaced for all English Local Authorities by the '1APP'. (Four copies, although some Local Authorities may request more for complex applications requiring a greater range of consultations. The duplicates can be photocopies of the original.) The form requires:
  - The name and address of the applicant and agent if applicable.

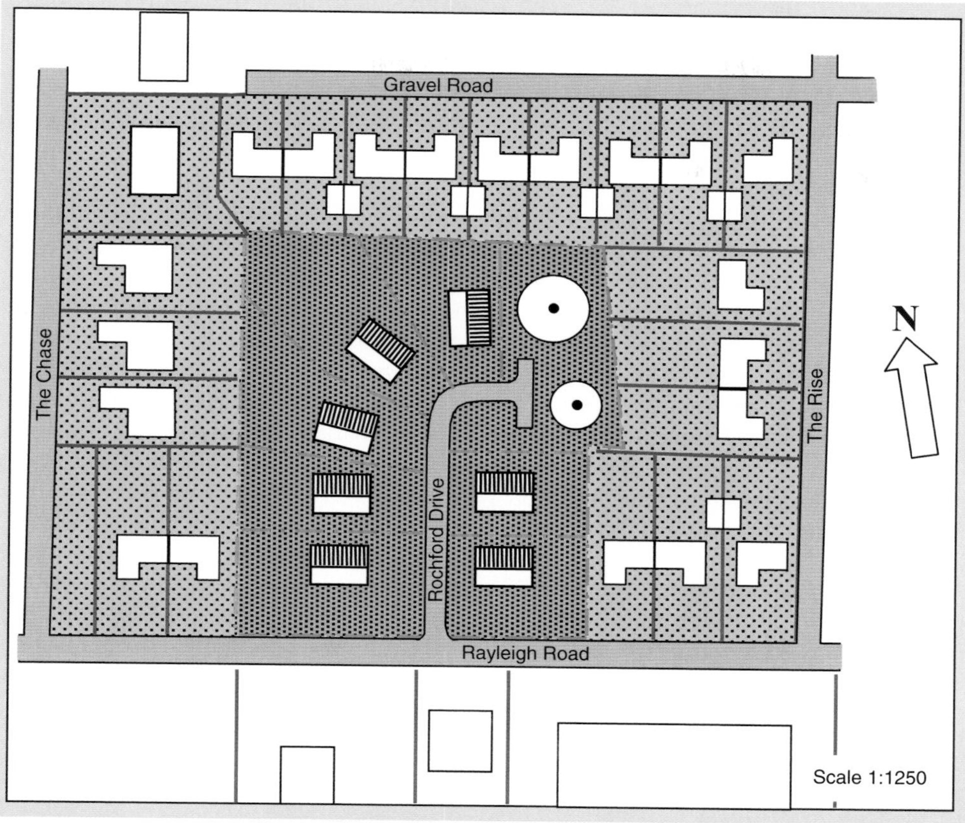

Figure 8.1 A scaled plot layout.

- Full postal address including postcode.
- Site area in hectares. (This is required to evaluate the density of the development. Where the site is Local Authority owned then a density of 50 dwellings per hectare (dph) is required. Under the PPS3 document a national indicative minimum density of 30 dph has been given with no upper limit. If a Local Authority wants to accept a density lower than 30 dph they must justify their decision under the requirements of paragraph 46 of the document.) A hectare is 10,000 m^2.
- A short description of the current use of the land and buildings. If the area is in a conservation area, on green belt or other restricted land, complete another appropriate form. (If the project had been commercial or industrial then relevant information would be required.)
- A short description of the proposed work including the use of the buildings and whether the development will have an impact on the public roads.
- What type of permission is required; outline, full, or reserved matters? Reserved matters are as a result of an outline application where

the LA planners placed stipulations before a full plans application could be considered, therefore the number of the outline planning application should be included.

- If protected species under the Wild Life Order 1985 are to be considered they should be listed. For example, if there is a pond on the site and crested newts are resident then it must be listed on the application.
- In this example the site is located near the driest part of the UK and water is a very important issue. On the application the source of water has to be noted: is it mains supply or private supply?
- Drainage for surface water and soil water must be identified. Applications for estates may overload an existing main therefore consideration by the utility companies will be given.
- Under Article 7 of the Town and Country Planning (General Development Procedure) Order 1995 a signed and dated statement of 'Certificate of Ownership' is required. Although it is not proof of ownership it is a legal declaration of ownership. The idea is to prevent a person applying for planning permission without the knowledge of the legal owner. There are exceptions such as where the legal owner is unknown, therefore advertisements must be placed for a minimum of 21 days prior to the intended application in local newspapers and A4-sized posters on or adjacent to the proposed site – see Figure 8.2. Before the requirement it was possible to apply for planning permission over land not owned by the applicant.

There are four alternatives:

- You are the owner or agent of the owner 21 days prior to the planning application.
- You are the trustee or intend to buy the property and have advised the owner of your intentions.
- You have tried to find the name and address of the legal owners by way of advertisement not less than 21 days prior to the planning application but have been unsuccessful. There may be several persons with a legal interest over the land, therefore those who have been identified would be noted on the certificate including the date that notice of your intentions was given.
- You have tried to find the name and address of the legal owners by way of advertisement not less than 21 days prior to the planning application but have been wholly unsuccessful. A declaration under Certificate D states the steps taken to find the legal owners including the name of the newspaper and the circulatory area where the site is situated. An advertisement requesting anyone who has a legal interest in the said property can be placed in the *London Gazette*. Originally called the *Oxford Gazette*, it is the oldest continually published newspaper, dating back to King Charles II. The original concept was to enable the Court of Charles II in 1665 to communicate with the masses about the Great Plaque; who England was at war with; and information about various campaigns

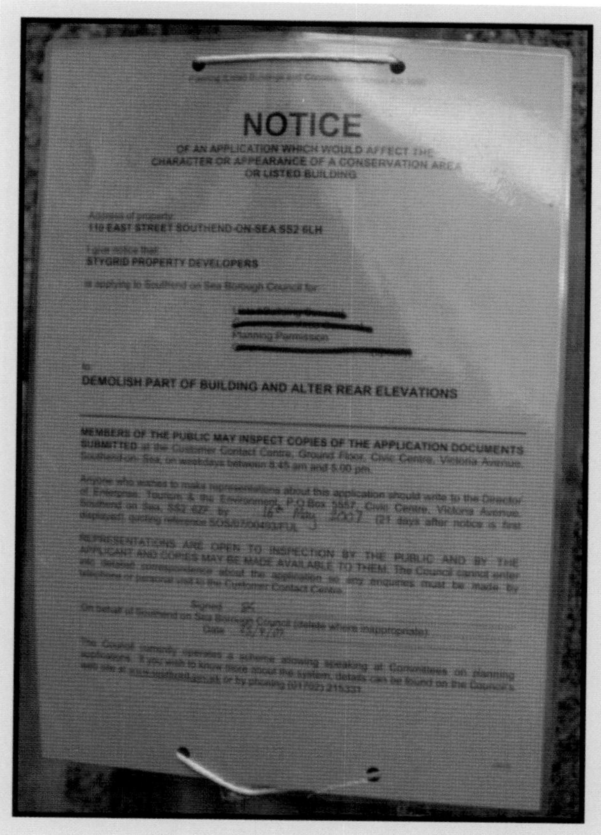

Figure 8.2 A notice of intention advertisement.

involving the armed forces. Today the *London Gazette* can be accessed via their website and it is still an ideal way of communicating information worldwide.

Certificate D is commonly used when the land has not been registered under the original Land Registration Act 1925 or the Settled Land Act 1925 (defining interests in land as possessionary title). Where land has been abandoned, made a gift without deed, or waste land for example bogland and unstable land it may be legally un-settled therefore not owned by deed. (All land is owned by the Crown or the Church of England – see Chapter 7.) Subsequently the Land Registration Act 1936, Land Charges Act 1972, Land Registration Act 1986 and 1997 and Land Registration Act 2002 have provided he opportunity and, since 1972, a legal obligation to declare any legal interest including charges (known as 'title') over land in England and Wales. Scotland and Northern Ireland have other legislation.

There is also a certificate for Agricultural Holding that must be completed with either form A or B. The declaration confirms that the planning application is not on land as part of an agricultural holding. If change of use is being applied for the tenants using the land on 21 days before the application must be notified. Their names and addresses should be confirmed on the declaration including the date they were notified. It is good practice when notifying tenants and owners that the letter be sent by recorded delivery as proof that the document had been sent.

In this example the two farm barns at one time would have been on agricultural land. However, the farm had been redesignated as building land many decades ago.

Planning form P1 can be downloaded from www.planningni.gov.uk/Devel_Control/Application_Forms/Forms.htm.

When the forms, drawings and other relevant information have been prepared they can either be posted or delivered by hand to the Local Authority. The procedure is shown in Figure 8.3. The forms have been standardised throughout England and Wales so that whichever Local Authority regulates the area, the forms and information required will be the same. Log on to any Local Authority website and download the standard 1APP form for planning applications pertaining to dwellings. There are a range of forms to cover other planning applications plus specific issues that Local Authorities may cover. For example, in parts of the West Country and Northern England and Scotland where granite is exposed, radon gas is an issue. In the East of England flooding is of concern.

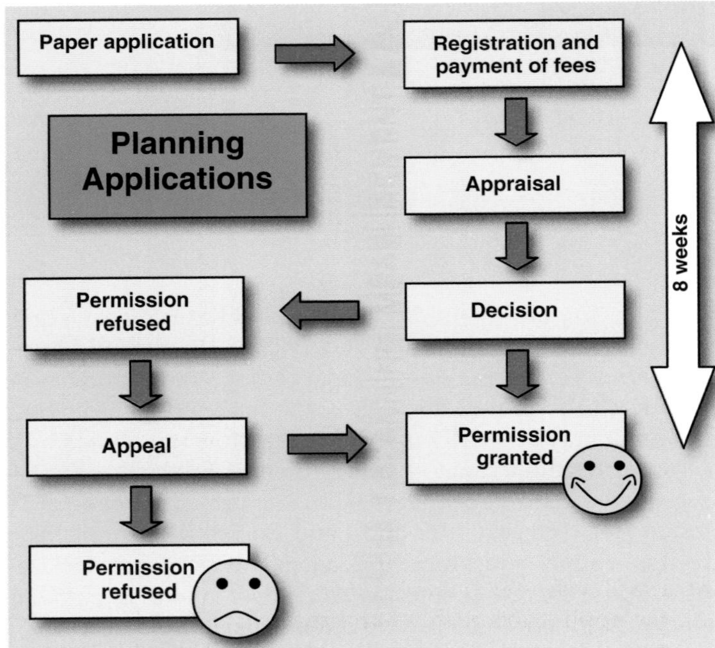

Figure 8.3 Planning Application flow chart.

Some Local Authorities have very old developments in their areas that have been designated conservation areas; therefore specific planning application forms are required. It is advisable to contact the Local Authority Planning Department and ask for advice before applying for permission.

8.2 The planning application procedure

Stage 1 [Receipt]

The process starts by presenting:

- the completed application forms (3 copies* of standard form P1)
- a signed certificate of ownership
- associated drawings (3 sets of copies* of each drawing)
- the correct fee.

* NB: Some Authorities may require up to 7 copies.
If any of the above are not presented, the application will be rejected.

In addition, include photographs or computer-generated impressions of what currently exists and the finished project as streetscene elevations, plus any relevant comments about vehicular issues including transport, for example if additional public transport will be part of the application.

Stage 2 (Registration)

The application will be checked to ensure all the documentation and the correct fee has been presented. The application will be assigned a number and passed to the person responsible for controlling the area where the proposed development will take place. He or she will notify the neighbouring property owners and where applicable any known societies that may have an interest in the development. The proposed project will appear in local papers and on the weekly register of applications that may be inspected at their offices or, in most cases, online. Most Local Authorities laminate a copy of the application on A4-sized paper to be displayed near or on the site (see Figure 8.4). The application is passed to the relevant statutory agencies to ensure it will not cause future problems. They include the entire utilities group such as gas, water, drainage, sanitation, electricity and cable companies, plus the highways and fire departments.

Stage 3 (Appraisal)

The planner will visit the proposed site and look at the surrounding area, comparing the application with what actually exists and ensuring the application is factually correct. In particular they will check whether there are any trees that may have a preservation order on them or whether the area has any

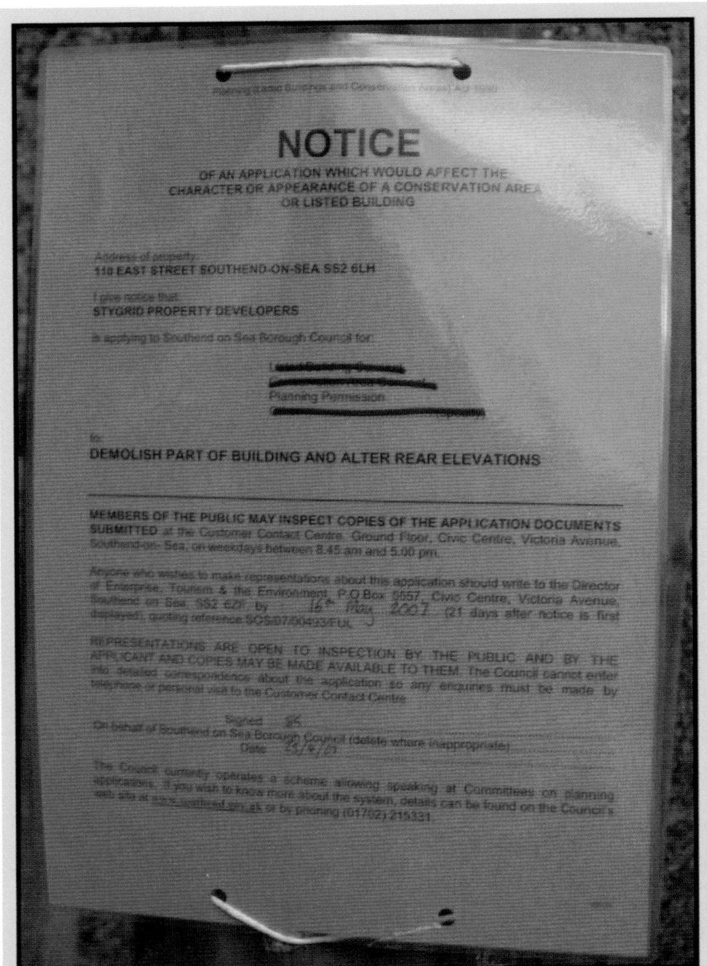

Figure 8.4 A notice of intention advertisement.

features of particular scientific or environmental interest such as a pond or evidence of badgers or other protected species of animal. There may be specific wild plants that have legal protection, although specialist organisations will know where most sites are and will be advised by the Local Authority of any planning or development applications. If it complies with the guidelines for the area set by the Local Authority such as use of land, type of and colour of the main materials and importantly whether there is sufficient off-street parking, then the planner may recommend that the development be granted. If, however, there are any objections from local neighbours or the utilities departments then the objections plus the planners' recommendations will go before a planning committee that normally meets on a monthly basis. The planner will also check on the history of the site to see if there have been any past applications for development.

Stage 4 (Decision)

There is a statutory eight-week allowance for the application to be considered. Many application forms ask whether extensions of time be considered to give the planners leeway if required beyond the eight weeks. This is more commonly used on complex proposals. If as previously mentioned the project is straightforward and complies with the Local Authority guidelines including off-street parking the planning officers will recommend the application is granted. If not, the Planning Committee will consider and debate the application in open proceedings. That means the members of the public and press may attend the meeting to observe who is for and against the development. The Planning Committee may visit the proposed site prior to the meeting to ensure a full understanding of the proposal.

If there are issues such as a protected tree or historic artefact on the site, the Planning Committee may approve the application with specific amendments or provisions such as protection of part of the site or, in the situation of the removal of a protected tree, that additional mature trees be planted elsewhere on the site to compensate for the loss. Developers and builders have been known to fill in ponds and fell protected trees prior to making planning applications. It is of little help after the event as the wildlife would have been destroyed before the Authority could do anything about the destruction and a fine may have a punitive effect on a large development proposal.

Where a planning application is refused, the applicant may appeal within six months of the refusal. The appeal is usually considered by an appointed person or persons working on behalf of the Planning Inspectorate. It may be in written form explaining why you consider the refusal unfair, or as an informal hearing more suited to larger or more complex projects in open forum, or where controversial projects have been refused a Public Inquiry.

In this example the developer has used the same local architectural practice as used on a previous development. That enabled the developer to negotiate a reduced fee as the new development used the same house types in a different layout. The developer and architect had a concept meeting to discuss the project proposal and entered into contract using the SFA 99 (Standard Form of Agreement document) available from the RIBA. The developer had previously commissioned a Town Planner to establish the feasibility stage and obtain outline planning permission, therefore the architect's contract was to obtain full planning permission and to provide the working drawings for Building Regulation Approval only. No site supervision or site meetings were required, as the developer and builder had experience of the house design from previous projects. The limitations of the required services were specified in the contract. Any additional advice or meetings would be based on additional fees.

8.3 Full planning permission

To apply for full planning permission more detail is required, so the form Application for Planning Permission. Town and Country Planning Act 1990

1APP should be used. Blank copies can be downloaded from any Local Authority website.

The form can be used for application to develop land for dwellings, industrial or commercial applications, and also where change of use may involve storage of hazardous substances or where waste management issues are concerned.

So far the developer has:

- bought the land including all property contained within the curtilage
- commissioned a solicitor to establish whether there are any easements, rights of way or covenants attached to any of the properties or adjacent land that may affect the development of the land
- commissioned a consultant Town Planner to establish whether the proposed housing development would receive planning approval. Due to there being two mature oak trees with preservation orders against them and a lower density of dwellings than the minimum 30dph, an Outline Planning Application was applied for
- received agreement in principle after 7 weeks from the Local Authority and been issued with Outline Planning Permission With Some Matters Reserved (PPWSMR). The location of the site and the existence of the protected oak trees enabled the lower density to be accepted
- asked a local architectural practice to apply for full planning permission and Build Regulation approval known as 'full plans' (not to be confused with Full Planning Permission). The architect had previously designed dwellings for the developer on another site. As copyright of the design remained the property of the architect, the developer negotiated a reduced design fee on this project. (The copyright of the design remains the property of the originator. The client had previously entered into contract for the design of a specific structure in a specific location therefore if the design is to be repeated elsewhere an additional fee is required from the copyright owner, the designer.) Additional work included drainage layouts, foundation design to suit the site and site layout plans.

The new forms required for planning require greater depth of detail therefore the architect had to produce a report under the heading 14. Biodiversity and Geological Conservation. The old barns adjacent to the mature oak trees had been the home for various wildlife, so a specialist was commissioned to report on the extent and location of the various wild life and fauna in the localised area. Other additional information included: 18. Residential Units (Including Conversion) where the number of existing dwellings that were to be demolished, the style of use and number of bedrooms in contrast with the new build expressed as a gain or loss of residential units overall.

8.4 Full plans

The term is used for confirmation that the building work will be in compliance with the Building Act 1985 and Building Regulations 2000 with amendments in 2002 and 2008. If the Local Authority is to oversee the compliance of the Building Regulations then an application on a form BC2 would be made and appropriate fees paid. The forms would be accompanied by sets of scaled detailed drawings to a scale of 1:5, 1:50, and 1:100 where applicable. The largest scale would be used to show greatest detail where the main foundation, drainage, floor plans, and roof layouts plus sections through would be 1:50 scale.

Full specifications should be included either on the drawings or as attached sheets. Some simple layouts may be produced to a scale of 1:100 with written specification. The plan layout should show relevant boundaries and service information; e.g. drainage for sanitation (soil/foul water) and surface water.

There is a statutory requirement that Full Plans Applications must be processed within five weeks from registration. However, Local Authorities ask whether an extension of time to two months may be agreed subject to the LA work loading. The permission is based on the regulations and amendments made and date of commencement at the time of the application. Building Regulations can be met using information set out in the Approved Documents. With modern regulations they are not retrospective therefore if a Full Plans Building Regulation Approval is given and an amendment comes into effect during the three years allowed before expiration of permission, the new amendment cannot be enforced. For example, the Regulations were amended with reference to the way heat loss calculations were to be assessed.

8.5 LANTAC

Where a developer has standard house types or intend to use a modular house type such as timber frame units, a reduced fee can be achieved where applications for Full Plans under Building Regulations is made. LANTAC stands for Local Authority National Type Approval Confederation. The design is scrutinised against the requirements of the Building Regulations (*nothing to do with planning issues*) and, if met, a LANTEC certificate is issued. The certificates are reviewed every three years or if new legislation comes into effect.

8.6 What are Building Regulations?

Building Regulations put simply are a set of legal requirements set against the health and safety of persons within the built environment. Prior to 1965 Local Authorities controlled the Health and Safety of structures and services with Acts of Parliament and bye-laws (see Chapter 7 for historic detail). In the opening paragraph of the Building Regulations 1972 it states:

> 'The Secretary of State for the Environment, in exercise of his powers under section 4 and 6 of the Public Health Act 1961 (a) and under sections 53, 61, 62, 64 and 90 of the Public Health Act 1936 (b), and section 24 of the Clean Air Act 1956 (c), each as amended by section 11 and of and

> Part III of Schedule 1 to the Public Health Act 1961, and all other powers enabling him in that behalf, after consultation with the Building Regulations Advisory Committee and such other bodies as appear to him to be representative of the interests concerned as required by section 9(3) of that Act, hereby makes the following regulations:-'
>
> (Crown copyright)

In 1965 for the first time a national document integrated the above legislation in an effort to standardise the implementation. However, the 1972 Building Regulations consolidated the 1965 Building Regulations and the amendments into a single volume of 188 pages. The Building Regulations applied to England and Wales apart from the Inner London Boroughs which were subject to their own London Building Acts. Although the Building Regulations are based on the Building Act 1984 (Building Regulations 2000), there are additional requirements when building in certain areas. There is a list of areas that have Royal Assent (permission from the Crown) that can enforce either extra regulation or specifically define building control. In 2003, enacted on 1 March 2004, the Secretary for State repealed six Local Acts, stating they were no longer needed. The reason for the variation may have been historic and peculiar to that Local Authority's area. The Building Control Officer dealing with the project will be able to advise of the requirements at the design stage. The Full Plans would be subject to all of the requirements therefore no additional variations would be considered after permission had been given.

8.7 Party Wall Act 1996

Before work starts on site the developer should contact the owners or tenants of the land bounded on all sides of the property where excavations will be within 3.00 m of the boundary. The letter should inform the neighbours of the intended works and offer to pay for a surveyor to look after the interests of the Party Wall Award. If both parties agree to using the same party wall surveyor, the instigator, in this case the developer, will pay for all fees and expenses. The party wall surveyors will carry out a survey of the property on and adjacent to the boundary noting any structural issues including damage. Photographs will be included of any issues not adequately noted by text alone. The date of commencement will be noted and the nature of the works. The surveyor(s) will inspect any applicable drawings and specifications to ensure that adequate protection of the neighbouring property can be maintained during the works. The Party Wall Act 1996 does not hold any powers as such and any redress to the instigator of the work will result in civil action through the courts.

8.8 Stages of building control

The Government have set six stages at which the building work must be inspected, termed 'Statutory Inspections'.

1. Commencement of works – the builder should advise the Local Authority 48 hours before the work will be proceeding. That enables the Building Control Officer (BCO) to be prepared to carry out inspections.
2. Excavation of foundations – 24 hours' notice is required prior to excavations. The BCO will want to:
   - inspect how deep the foundations will be taken
   - ensure the sides of the trench are plumb and regular
   - ensure there are no exposed tree roots entering the excavation
   - ensure there is no made up ground that will be weaker than the existing sub-soil
   - walk in the foundation trench to ensure there are no soft spots. If there are, the most common remedy is to excavate deeper in the soft area and step the foundation. If this is not possible or practical, a structural engineer will be needed to design an alternative foundation detail and agree it with the BCO before the foundation concrete is placed.

   It is not acceptable to concrete in the foundations before the BCO has signed them off. BCOs will require the foundations to be broken out and reinspected as he or she will be signing off that they have carried out the inspection. If there is steel reinforcement to be included in the foundation the BCO will also want to inspect it prior to concreting in.

   The BCO will also want to know who is producing the concrete for the foundations. If it is a reputable ready mix company then generally there should be no problem with the quality of the concrete. On this site a local firm of groundworkers had been used and an established ready mix concrete supplier. The BCO inspection was carried out as the first site visit of the day and the concrete had been ordered for late morning. If the work had been carried out by an unknown contractor and the concrete produced by one of the 'mix and lay' contractors the BCO may visit the site during the work to ensure the concrete is of reasonable quality. This applies more to house extensions though. Some of the 'mix and lay' companies have a reputation for quoting a cheap price for the job and then at the end stating they had used more concrete than quoted for, thus increasing the costs in some cases very significantly. If the concrete is of the required quality it is something for the client to sort out as it is not a quality issue. Other companies get a reputation for not using good quality aggregates and less cement than required, again cheating the customer.

   The BCO is the professional person inspecting the foundation, not the quality control manager. However, the Building Regulations state that the materials used must be of a minimum quality and as such the BCO can insist on proof that the materials, in this case fresh concrete, meet the specification. Hopefully most of the 'cowboy' contractors have ceased trading, though they emerge as a different company from time to time. Generally a company that only has a mobile phone number and name on the side of their vehicle would give cause for concern. Reputable companies would have landline telephone numbers and an address that could be checked out. BCOs and Trading Standards Officers will know of the rogue companies, but with several tonnes of set concrete in the ground, is it worth taking the risk?

3 Underbed for oversite – where a solid ground floor has been chosen, the walls up to dpc level would be erected on the foundation. The foundation may be a strip, or trench fill (see Cooke (2007), Chapter 4 for more about foundations). The ground floor area would then be covered with consolidated hardcore, which could be broken bricks or demolition concrete crushed to about the size of a fist. The hardcore should be well compacted and level, ready to receive a layer of ash or sand, known as 'blinding'. Where gravel pits are local, a material that contains some clay but is mostly fine aggregates known as 'hoggin' can be used. The material will fill the voids left between the hardcore and provide a level and smooth bed 50–75 mm thick onto which the damp proof membrane (DPM) can be placed. The DPM minimum thickness is 1000 gauge (250 μm) polythene and where possible should be in one sheet (see Figure 8.5). If joins are required, the overlap should be in compliance with the manufacturer's instructions as detailed in the British Board of Agrément (BBA) certificate (see Figure 8.6). All materials and products must be tested and approved by an independent testing company. A leading manufacturer of damp proof membrane material is Visqueen. Their product information is available on the website and an electronic version of their BBA certificate available on request, allowing confidence that the product/material is of a set quality and detailing how it should be used. In contrast, other lesser quality polythene sheeting may look similar, but if the BCO is not satisfied with the quality it may be difficult trying to obtain a BBA certificate for quality assurance and correct use. Alternative certification can be provided by proof of compliance with British Standards or similar codes from any member state of the European Community. The Building Regulation number 7 states that:

> **'μm'**
>
> **μ is an ancient Greek letter 'm' and pronounced 'mu'. It is used to represent 1,000,000th (10^{-6}) or a micro, therefore:**
>
> **μm = 1,000,000 of a metre**
> **μm = 1,000 of a millimetre.**
>
> **250μm = (1 / 1000) x 250 = 0.25mm**
> **300μm = (1 / 1000) x 300 = 0.30mm**

Figure 8.5 Still using ancient Greek letters.

'Building work shall be carried out –

(a) with adequate and proper materials which –
  (i) are appropriate for the circumstances in which they are used,
  (ii) are adequately mixed and prepared, and
  (iii) are applied, used or fixed so as adequately to perform the functions for which they are designed; and

(b) in a workmanlike manner.

Crown copyright: 2000. Statutory Instrument 2000 No. 2531.

In recent years concern over radon gas in parts of Cornwall and Somerset, and parts of the North of England and Scotland mean that the

Designated by Government to issue European Technical Approvals

CI/SfB

(13.9) n6

Visqueen Building Products
Maerdy Industrial Estate
Rhymney
Tredegar
Gwent NP22 5PY
Tel: 01685 840672 Fax: 01685 842580

Agrément Certificate
No 94/3009
*Third issue**

VISQUEEN BUILDING PRODUCTS DAMP-PROOF MEMBRANE

Membrane étanche à l'humidité

Feuchtigkeitssperre

Product

- *THIS CERTIFICATE REPLACES CERTIFICATE No 87/1846 AND RELATES TO VISQUEEN BUILDING PRODUCTS DAMP-PROOF MEMBRANE, A LOW-DENSITY POLYETHYLENE MEMBRANE FOR USE IN SOLID CONCRETE GROUND FLOORS NOT SUBJECT TO HYDROSTATIC PRESSURE, TO PROTECT BULIDINGS AGAINST MOISTURE FROM THE GROUND.*
- *The product is available in thicknesses of 250 μm, 300 μm and 500 μm.*
- *It is essential that the product is laid in accordance with the recommendations of clause 11 of CP 102 : 1973 or with this Certificate.*

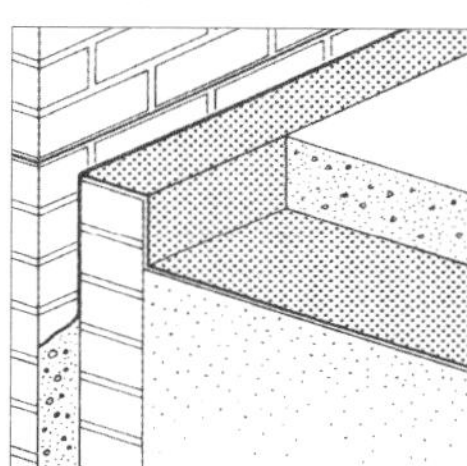

Regulations

1 The Building Regulations 2000 (as amended) (England and Wales)

The Secretary of State has agreed with the British Board of Agrément the aspects of performance to be used by the BBA in assessing the compliance of damp-proof membranes with the Building Regulations. Inthe opinion of the BBA, Visqueen Building Products Damp-proof Membrane, if used in accordance with the provisions of this Certificate, will meet or contribute to meeting the relevant requirements.

Requirement:	C4	Resistance to weather and ground moisture
Comment:		The product will meet this Requirement. See sections 8.1 and 8.2 of this Certificate.
Requirement:	Regulation 7	Materials and workmanship
Comment:		The product is an acceptable material. See section 13.1 of this Certificate.

2 The Building Standards (Scotland) Regulations 1990 (as amended)

In the opinion of the BBA, Visqueen Building Products Damp-proof Membrane, if used in accordance with the provisions of this Certificate, will satisfy or contribute to satisfying the various Regulations and related Technical Standards as listed below.

Regulation:	10	Fitness of materials
Standards:	B2.1 and B2.2	Selection and use of materials, fittings, and components, and workmanship
Comment:		The product complies with these Standards. See section 13.1 of this Certificate.
Regulation:	17	Resistance to moisture
Standard:	G2.6	Preparation of a site and resistance to moisture from ground —Resistance to moisture from the ground
Comment:		The product can enable a floor to satisfy the requirements of this Standard. See sections 8.1 and 8.2 of this Certificate.

3 The Building Regulations (Northern Ireland) 2000

In the opinion of the BBA, Visqueen Building Products Damp-proof Membrane, if used in accordance with the provisions of this Certificate, will satisfy or contribute to satisfying the various Building Regulations as listed below.

Regulation:	B2	Fitness of materials and workmanship
Comment:		The product is an acceptable material. See section 13.1 of this Certificate.
Regulation:	C4	Resistance to ground moisture and weather
Comment:		The product can enable a floor to satisfy the requirements of this Regulation. See sections 8.1 and 8.2 of this Certificate.

4 Construction (Design and Management) Regulations 1994 (as amended) Construction (Design and Management) Regulations (Northern Ireland) 1995 (as amended)

Information in this Certificate may assist the client, planning supervisor, designer and contractors to address their obligations under these Regulations.

See section: *5 Description* (5.1).

Readers are advised to check the validity of this Certificate by either referring to the BBA's website (www.bbacerts.co.uk) or contacting the BBA direct (Telephone Hotline 01923 665400).

Figure 8.6 British Board of Agrément Certificate.

DPM must be suitable for use in radon gas conditions. In other areas where landfill sites are in close proximity or the building land has been reclaimed boglands, or marshes methane gas would be the issue. *Not all DPM materials are the same.* If the designer has specified a product and/or a manufacturer then it should be used. It is good practice for the designer or specifier to add a statement that the materials should be as specified or as equally approved by the designer. In the event that the building contractor cannot obtain the specified material then permission/agreement must be sought with the designer before a substitute is used. There may be a scientific reason for choosing a specific product as suggested above.

If a concrete oversite is to be laid directly on the DPM then floor insulation will be placed on top of the concrete. Alternatively the floor grade insulation of the correct thickness and grade should be correctly placed on the DPM prior to the slab being placed. If a plastic sheet membrane is not used, an alternative DPM could be three coats of water-based cold bituminous paint, or hot applied asphalt screed or tanking, or vapour-resistant waterproof adhesive and quarry tiles directly on the concrete oversite. The BCO will be interested in the detail where the floor meets the walls to ensure there will not be a breach of the moisture barrier.

On this site a suspended block and beam floor design had been used therefore the inspection for 'Underbed for Oversite' was not required.

4 Damp proof course – the concrete block and beam ground floor is supported by the walling on the foundation terminated at dpc level. To prevent the passage of moisture entering the ground floor (the objective of the inspection) a continuous damp proof course should be bedded in suitable grade mortar and lapped at all corners and joins. The ground below the floor must be a minimum of 75 mm below the lowest edge of a ground plate but generally at least 150 mm below the suspended floor to prevent dampness rising and allow for any gases to be ventilated out. The supporting walls must have suitable wall vents to allow cross ventilation to take place in compliance with the Building Act 1984 – Building Regulation 2000.

(i) Loading A1
(ii) Ground movement A2
(iii) Preparation of site C1
(iv) Dangerous and offensive substances C2
(v) Subsoil drainage C3 (a) the passage of ground moisture to the interior of the building. (b) damage to the fabric of the building – (this part of the regulation is particularly relevant to timber framed buildings)
(vi) Resistance to weather and ground moisture C4

The BCO has the authority to inspect and be satisfied that the work complies with all of the above sections. Section 2 of the Building Regulation 2000 states that:

'A person carrying out building work shall not –

(a) cover up any excavation for a foundation, any foundation, any damp proof course or any concrete or other material laid over a site; or

(b) cover up in any way drain or sewer to which these Regulations apply, unless he has given the Local Authority notice that he intends to commence that work, and at least one day has elapsed since the end of the day on which he gave notice.

The site in the example had a specialist supplier of suspended block and beam floors to supply and fit the floors. The BCO was able to see the dpc beneath the concrete beams prior to the specialist contractor bedding in beneath the edge floor blocks.

5 Drains (FW or SW) – drains are another health-related issue. Before the drainage runs are backfilled over or commissioned they will be inspected for fall and water tightness. The normal method of testing would be for the groundworker to use a drain 'bung' or steel plug at the outfall manhole or inspection chamber. (This is a rubber ring sandwiched between two steel plates with a wing nut and bolt that when tightened forces the rubber ring to expand. The bung prevents any air leaking through the outlet.) A second bung with a tube allowing a small hand pump and a manometer to be connected is fixed at the other end of the drainage run. When secure, the drainpipe has the air inside pressurised and noted on the manometer. For example, the 100 mm diameter combined sewer/drain had to maintain 1kPa air pressure with an allowable drop of 0.25kPa over a 5 minute duration. It is normal to put the pressure about 10% over to allow for an initial drop and then time it. If the test fails it may that the bung is not tight enough, so it is adjusted and the test repeated. In colder weather the air pressure will fall if the temperature drops during the test. Just 1°C will cause 38 mm water gauge drop. The drainage run will be visually inspected and if acceptable permission to backfill will be given. If it is not possible to achieve a satisfactory air pressure result then a water test should be carried out.

Like foundations the BCO must see the stage and not just take someone's word the work has been carried out. It is good practice to temporarily bung both ends of the drainage runs before any backfilling to prevent any debris becoming stuck in the pipes. Inspection chambers may have temporary covers over them so materials may fall into the chambers.

Just before the completion and hand over takes place the drains should be inspected again and a water test carried out. In a similar way to bunging the drain for the air test the water test requires the drain to be filled with clean water and a short length of pipe known as a header pipe containing a head of water is fitted to the higher end of the drainage run. The water test should take place over 30 minutes after a permitted permissible top up has been carried out. Air trapped in the pipework or initial absorbency of the pipe material will show a drop in head.

6 Completion –when applying for Full Plans permission (nothing to do with planning permission, it is the term for Building Regulation permission only), on smaller projects the applicant can request that the BCO issues a completion certificate signing off the completed work confirming compliance with the Building Regulations. Where a project has an

architect or surveyor as the lead consultant he or she as part of the contract will issue a completion certificate. If the project is governed by a Client Management Agreement the principal contractor and designer will jointly issue a document stating that the work has been carried out in compliance with the Building Regulations and design documents. The certificate will be required by the insurance company who will eventually insure the building as confirmation that the works meet the minimum standards of the Building Regulations including fire safety (Section B1 to B5 of the Regulations)

Additional inspections are also carried out, however they are not statutory:

- Floor/roof joists – some Local Authorities also inspect the size, the centres for the joists and strength of timber used before they are boxed in with the ceiling materials. Joists often require higher grades of timber so the BCO will look for the grading marked on the joists.
- Roof carcass – for similar reasons to the joist inspection the BCO will want to check the grade of timber used, centres of the trusses or rafters and to check, if gang nailed trusses have been used, that the joints are still correctly held together.
- Other inspections – such as those required for timber-framed buildings will be identified by the BCO.
- The BCO will check that the correct materials have been used for insulation purposes both thermal, and the elimination of cold bridges, and sound insulation. The BCO may request that the building be tested to prove compliance with E1–E3 of the Regulations.

So far we have:

- The project being inspected by the Local Authority Building Control Department under the Building Act 1984 and Building Regulations 2000 plus amendments.
- The BCO has carried out a series of statutory inspections related to health and safety of the people who will eventually use the building and the protection of the ground water and public health of people in the area (more specifically the requirements of the foul water sewer).
- In April 2006 Part L of the Building Regulations was amended to include the air tightness of dwellings. There are two categories for testing:

  1. Dwellings that have adopted approved construction details
  2. Dwellings that have NOT adopted approved construction details.

As the design of the dwellings for this site had been completed for a previous development prior to the Regulation change they had not incorporated the 'approved construction details'. To change the design would require considerable work by the architect therefore the developer opted for on-site testing. Part L1A requires two tests on each dwelling type for developments of more than 4 dwellings and less than 40 dwellings. The developer had decided to use only two different designs on the development and commissioned a specialist air testing consultancy to advise and test during construction. That way there was a very high chance that the final test would be successful.

The tests included porosity of the masonry inner leaf blockwork, seals on the loft hatch, seals around the light ceiling roses and power sockets. On the final test when the house was complete the extractors in the kitchen, bathroom, en-suite and WC were taped up and the windows and external doors shut. Even the overflows and waste pipes to the sink, bath, basin and WC were sealed. All internal doors were left open allowing the free passage of air. The entrance doorway had a sealed blank door unit incorporating an electric fan fitted in the opening.

The technician monitored the speed of the fan pushing the air into the building from outside, increasing the air pressure to 50Pa. Calculations are made electronically comparing the pressure applied by the fan with the drop in pressure from the building envelope. The normal air discharge rate for a dwelling up to 450 m^2 floor area is 9.0 $m^3/hr/m^2$ at 50Pa with a best target figure of 3.0 $m^3/hr/m^2$ at 50Pa. There are several factors that have to be considered such as the ambient temperature within the building, wind pressure and direction, and outside temperature.

When the test has been successfully completed a certificate is issued.

When the dwellings are finished and the certificate of completion issued, the contractor will review the work with the initial buyer or agent of the client for a process known as 'snagging'. Items such as damaged facing bricks, marks on paintwork, poor quality of paint finish, scratched kitchen counter tops etc. have nothing to do with the completion certificate as they are cosmetic. The contractor has an obligation to rectify any issues before final handover. There is also a defects liability period which starts when the lead consultant (in this case the architect) has issued the certificate. The defects liability period is set out in the contract, although six months is commonly accepted. The liability covers such things as shrinkage of materials and items like door catches not working properly and needing adjusting. It does not extend to poor design, though, which is the domain of the designer.

Many house building developers also become members of the National House Building Council (NHBC). When selling new houses the buyer and money lender/mortgage company will normally require the dwelling to have additional insurance. To obtain insurance warranty under the NHBC scheme the design details must comply with the NHBC requirements, which are in addition to the Building Regulations. Currently the Building Regulation Monitoring can be carried out by an Approved Inspector, so instead of using the Local Authority BCO the inspections and certification can be carried out by the NHBC, who are one of the licensed Approved Inspectors.

The insurance warranty policy given by the NHBC does not come into effect until two years after the completion certificate and start of the NHBC warranty. In the first two years from the date of the insurance policy any defects must be addressed to the builder who carried out the work. These would be classed as patent defects, easily seen by the client or professional advisors. After the initial two year period has expired any defects (known as latent defects – those that were not visible for several years) that are subsequently discovered would be addressed to the NHBC insurance claims department, who hold cover for another eight year period.

Major defects such as foundation problems and subsidence that become apparent after the two year builder's insurance should be dealt with under the NHBC insurance warranty policy. As with other insurance warranty policies there is a measure of what exactly is covered and on whom the burden of proof is placed. For example, what actually caused the failure? In practice it is normally better to instruct a qualified Building Surveyor or qualified Structural Engineer to comment on the failure as there have been cases where insurance companies will not meet their obligations unless forced to. There are several insurance companies that offer building developers separate insurance to cover as a warranty against latent defect claims up to 10 years from the date of completion.

As mentioned, the builder is responsible for any defects that are noticeable in the first two years. Defects could be such things as minor or superficial cracks due to drying out of materials such as concrete blockwork, timber joists and the like. It should be noted that the NHBC state on their website that the builder is not responsible for 'normal shrinkage' due to the building drying out. That is very debatable. If the builder has used materials in accordance to the manufacturer's specifications and constructed in accordance with the relevant guidance of the British Standard Codes of Practice then there should not be any defects such as cracking. However, many builders do not meet the set standards of good building practice.

For example, timber should have a specific moisture content before it is used. Much of the timber is supplied in a saturated state and not left to dry out. The consequence is the timber will shrink when the building's first owners move in. Timber joists should have a moisture content of less than 12%MC. The timber should be delivered on a sheeted lorry to prevent it becoming wet during transit and then stacked on site clear of the ground, 'sticked' and covered with a waterproof sheet with allowance for ventilation. Many builders accept timber that is so wet that when stood on end water actually puddles from the timber. The carpenters cut the floor joist to length and the bricklayers build in as work proceeds. The cavity wall has insulation built in thus the timber has virtually a sealed set of tracheids (Tracheids are drinking straw-like tubes that enabled the tree when it was growing to transport water through it – the 'grain'.) The plasterboard ceiling is then tacked beneath the floor joists, scrimmed and skimmed further, preventing the timber from drying. When the first occupants move in, it is very common for the central heating pipes within the floor void to heat and dry the timber out, thus drying shrinkage manifesting as cracked ceilings or movement of covings. On the floor deck above the signs of significant drying shrinkage include skirting boards that have large gaps between them and the floor deck. Where the builder has tried to keep the timber dry and allowed it to stand prior to being used, there will be some very minor shrinkage. If that is the case most good builders will have the work remedied. I have seen cracks in walls that have been 1–2 mm wide that the builders have put down to normal drying shrinkage. They are not 'normal' and should not be accepted.

Another common 'oh that's normal drying shrinkage' is cracks in lightweight and aircrete blockwork walls. If the blocks have been used whilst they have a high moisture content (when delivered hot or 'green' they should be restacked to allow to dry out), they will shrink longitudinally. The manufacturers produce excellent technical information about drying shrinkage and how to prevent it. The British Standard Code of Practice recommends that masonry reinforcement be used in areas of high stress such as the spandrel panel below and above window openings. (For further technical information see Cooke (2007), Chapter 5.)

Movement joints should be used where block runnages exceed 6.00 m in length or 3.00 m from a return. Whole blocks should be used where the block is to become a bearing block. Where not practical then a minimum of two thirds should be used unless the structural engineers instructs otherwise. As you can see, buildings will not crack unless there is a reason; it is not acceptable to state that the cause is 'drying shrinkage'. Good building developers will require traditionally built masonry buildings to be ventilated for several weeks before they are handed over. They provide guidance to the new owner about the likelihood of condensation being produced, especially in the colder months, if the dwelling is not adequately ventilated, and that the central heating should not be turned up to full working temperatures until the building has a reduced moisture level. In practice it takes about a year for the moisture content to become stable throughout the dwelling.

Dwellings may be tested as part of the Building Regulation 2000 Approved Document L1a 2006 amendment and comply. If 'shrinkage cracks' are inevitable as suggested by the NHBC, will the building no longer comply?

The builder, and in this case the developer, will provide the new owner of the dwelling with a building manual or log book. It comprises instructions on:

- how to use any fixed appliances including central heating boilers, thermostats, extractor systems etc.
- the 'customer instructions' that are supplied with larger appliances such as central heating boilers
- certificates that all gas appliance have been checked and certified by a CORGI registered person
- certificates to confirm the glazing has been installed and certified by a FENSA registered contractor
- an Electrical Installation Certificate issued by a contractor who is a member of the Institute of Electrical Engineers (IEE) or similar recognised qualification to be in compliance with BS 7671: 2001. The certificate confirms that any electricity cabling and electrical switching gear work has been carried out and checked in compliance with the Electrical Regulations and Approved Documents P 2006.

Chapter 9

Shop refit

Now we have looked at the building of a small housing estate this chapter follows the project of a hairdresser opening a new salon.

This chapter has been based on a real project, but parts have been modified to provide detail of how a larger shopfitting company would have handled the project. (In particular the shopfitter in the real project was a small firm basically comprising two partners who employed various tradesman as required.)

John has a men's hairdressing salon on a busy main road with no parking area or street parking. He and his three staff have to park their cars several streets away if there are any spaces, as do most of the clients. After ten years the time had come to consider moving to other premises where he and his clients could park relatively easily. John had several options:

1 Sell his freehold shop and flat above and buy new premises.
2 Sell his existing shop and flat and rent new premises.
3 Lease his existing shop and continue to lease the flat above and lease new premises.

After careful consideration and discussion with his accountant John decided to sell the flat and lock up shop and buy a new shop.

At this stage John arranged for three local estate agents to value his existing premises and send him details of suitable lock up shops in the local area. After viewing several premises John decided to proceed with buying the freehold on a local lock up shop and put in an offer. The shop had previously been used by a florist therefore considerable work would be needed to convert it to a salon.

John has been in business for over 35 years and knows what he wants. However, his expertise is in hairdressing and not construction. Oe of his clients recommended a local shopfitting company to carry out the full transformation. A meeting was arranged on site. As the shop unit was empty John requested the keys from the estate agent and arranged to meet a director from the shopfitting firm on site. After discussion about what John wanted and what could be achieved, the next stage was for the shopfitter to provide a guide price for the work. Approximate overall dimensions were taken of the existing floor areas, storey height and frontage. Note that at this stage there was no contract between John and the shopfitter therefore either party could stop at any time without penalty.

The shopfitter provided a written guide price for the work based on the timescale John wanted and the approximate budget for the work. Unfortunately the total sum came in higher than John had envisaged therefore more consideration had to take place such as could the finishes be cheaper or would the overall image suffer? With some changes to the plan John decided to instruct his solicitor to proceed with buying the shop unit. Ownership or being an agent of the owner is an essential part of Planning and Building Regulations Approval. Also, the shopfitter requires evidence of ownership prior to carrying out any work on the premises.

John was in a good financial position but could not afford to buy the new premises before selling his shop and flat. He applied to his bank for a commercial bridging loan. Bridging loans are short-term loans that are normally at a higher rate than conventional long-term loans. That way he could continue trading in his existing shop whilst the new premises were being converted. Whilst the work was being carried out he could put his shop and flat on the market and specify a completion date after he intended to move to his new salon. In this particular scenario John and the shopfitter had a letter of confirmation but no formal contract. He trusted the local shopfitter, who had been recommended by one of his own clients. However, the trust could have been misplaced. In this case everything went through as planned but it could just as easily have gone sour.

The shopfitter prepared a lump sum for the whole project based on unit costs. They priced for taking the project from the design stage through to completion. Using computer aided software the shopfitter produced an electronic impression of what the finished salon would look like and with a few minor changes John accepted the design. Other than the overall dimensions taken at the initial meeting all other dimensions were approximate therefore a measured survey was needed.

9.1 The project

As the use of the premises had already been a shop there was no need for a change of use application. The shopfront needed replacing and a new backlit fluorescent lighting sign box above the shopfront plus shop awning and blind therefore a Local Planning Application under the Town and Country Planning Act 1990 would be needed. Considerations such as colour of the proposed sign and position to the road and any road junctions and street furniture will be taken into account of the application. For example if the proposed sign was predominately either red or green and there were traffic signals directly in front of the sign it could cause a problem therefore restrictions would be imposed on the planning approval. In this case there are no traffic signals or other street furniture to be considered. The shop blind and awning would not exceed the shop curtilage. Although there is a public footpath approximately 2.0 m wide alongside the road the title deeds of the property show that an area outside the shop is still private land and part of the property. The reason for the privately owned area is to enable the shopkeeper to display products or wares outside the shop (see Figure 9.1). Many buildings including the prestigious 'Gherkin' have a series

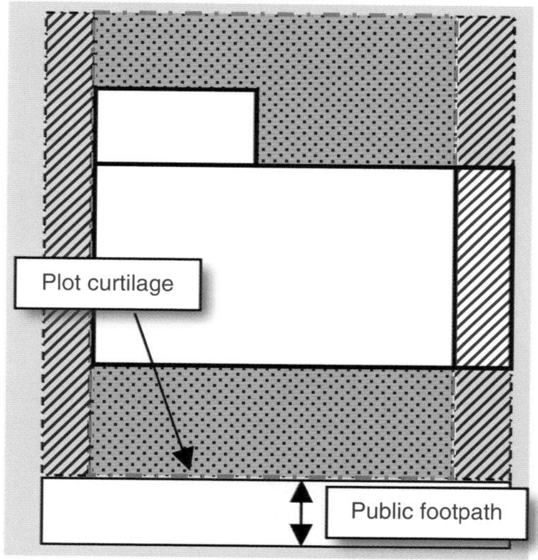

Figure 9.1 Plan showing plot cartilage.

of metal studs on the paved area surrounding the structure identifying where the public area finishes and the private area starts. The line is the boundary also termed curtilage and the owners can dictate the conditions of entry onto their land. For example no photography may take place without specific permission anywhere within the curtilage. So if you want to photograph the magnificent structure it must be from the public area only.

9.2 Planning application

The shopfront of the existing premise was an old wooden single-glazed front with a wooden single-glazed door with louvre blades in the fanlight above. Ideal for the florist, as a cool shop is essential to keep the flowers fresh. However, it would not be suitable for the new salon. Although the new shopfront would not increase the size of the structure it would be a material change and require permission (see Figures 9.2 and 9.3).

So far:

- Where a larger shopfitting company has been given the work a formal contract would be written up. The JCT Minor Works 2005 (JCT:MW.05) (see Chapter 1). The Company Secretary would sign on behalf of the shopfitters and the client would sign to confirm agreement to the contract.
- The Company Secretary is normally a director with a professional financial background, perhaps with accountancy training. He or she may also be the finance director and in that case the following description would

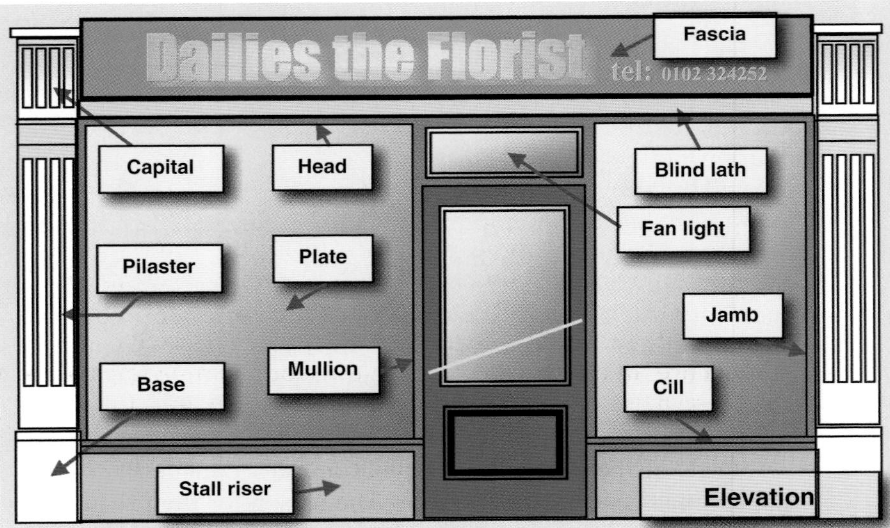

Figure 9.2 Front elevation of the shop front.

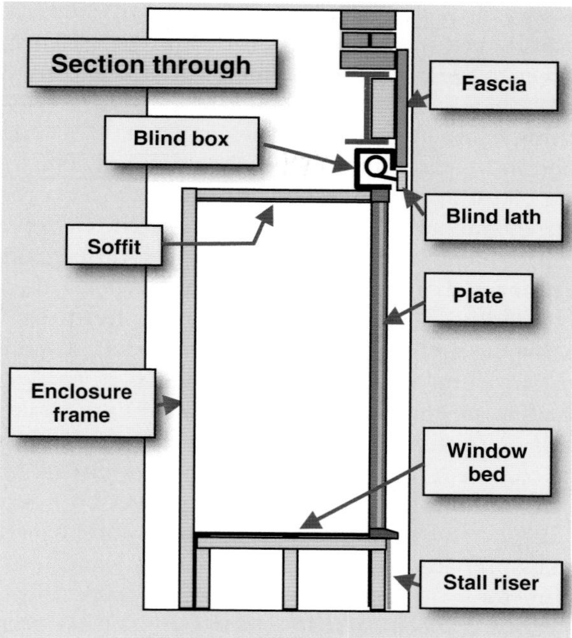

Figure 9.3 Section through of the shop front.

apply. They will have responsibility for the 'offices' (the paperwork side of the company) and control the finances of the company. It is most likely that they will have no construction knowledge. They will head the financial part of the company, including the accounts department, the wages department and human resources. The legal aspect to trading would also fall into their remit: company registration, preparation of the tax returns for the Inland Revenue, Public Liability insurance, contract insurance and basically taking responsibility for anything legal that involves the company.

- In contrast, the contracts director of the shopfitters would have a construction background and responsibility for the construction side of the company. He or she would control the works (a term used for the workforce other than those under the finance director). Health and safety for the company would normally be under the contracts director and in those famous words 'the buck stops here': they would be legally responsible for any accident that was proven to be due to their own negligence. In an extreme case the director could be imprisoned if it were proven that the correct procedures were not in place for health and safety. All risk assessments and company procedures known as 'method statements' should be sent through to be checked off against the overall safety plan.

The contracts director in this case is also the representative responsible for obtaining work for the company. (Some larger shopfitting companies have several representatives seeking work.) At this stage the director has completed his part by getting the enquiry for the project. He now passes the paperwork over to the designer and the estimator to provide an approximate costing.

- In larger shopfitting companies the designer, as the name suggests, converts the sketches and written notes from the contracts director or representative into a pictorial presentation. Until recently, the presentation would have been several hand-drawn isometric sketches painted in with watercolours to provide an artist's impression. The designer had to be an excellent artist more than a shopfitter. Today, however, the designer needs to be a computer operator, as the format is electronic. Software programmes enable the designer to draw in plan layout, convert into elevations and then 3D/isometric. Colour rendering can be easily changed, enabling the client to see alternative options almost immediately, something the hand-drawn artist's impressions could never achieve. Other software such as 'Arcon 3D Architect store' provides virtual reality, enabling the client to see the proposed finished project on a screen. The company Eleco have a short online video demonstrating exactly how a presentation can be prepared – see http://store.eleco.com.
- The contracts director has another meeting with John (the client) to show him the proposed electronic video artist impression prepared by the designer, samples of materials and finishes including sections of the shopfront framing and stone counter tops. The client has agreed to the specification and the shopfitter estimator then prepares a quotation for the work (see Figure 9.4).

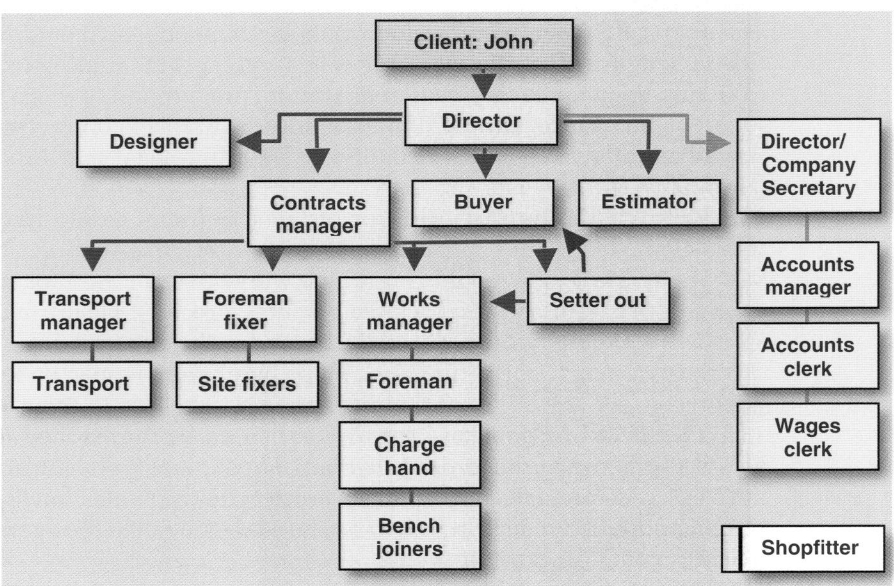

Figure 9.4 Shopfitting team and the client.

The order has been secured based on the quotation. (NB: A quotation is an actual sum set against specified work, in contrast to an estimate, which is an approximate sum that can be more or less when the final bill is presented.)

- Preparing an estimate for fitting out a shop is unlike most other estimating. For example, counters and desks are usually individually designed and made for each project therefore to calculate an exact cost would require a full take off of material and estimated labour hours/costs. Apart from taking a huge amount of time and resources, it may all be abortive if the company does not win the contract. The experienced estimator will have an approximate figure based on counters that the company have made in the past. Additional costs for special materials such as a stone counter top would be either based on a cubic 25 mm (the old cubic inch) or a quotation from the stone masonry.

  The estimator normally goes out to different suppliers to obtain prices for specialist materials and sub-contract work. Rarely now does a shopfitter directly employ electricians, ceiling fixers or plumbers etc., as to pay their wages when they are not working on site would be cost-crippling. Enquiries are sent to the various specialists and suppliers inviting them to quote for specific parts of the project. In smaller projects such as the salon fit-out the ceiling contractor would attend site to take working dimensions. On larger projects a full set of working drawings would have been prepared by the lead consultant and a copy would be included with the enquiry. It is very important to emphasise on the letter addressed to the specialist that it is an 'enquiry' and not an 'order'.
- The whole project is now passed to the contracts manager, who will form the team to work on the job. The contracts manager essentially is a hub for each project team to go to. In the shopfitters I worked for there were about eight setters out, each working on at least one project. All contracts should have a specific time element; when to start and a completion date. The contracts manager will produce a contract plan based on the two specific time elements. The project would be handed over to a setter out to organise the supply of materials for the project.
- Setters out would convert the designer's idea into working dimensions and materials lists. In shopfitting virtually everything is drawn out full size. However, where materials are the same for long runs, break lines would be drawn in. The overall dimension would be shown and the deduction would be written in as 'add xyz'. Then when the rod is being used for, say, the shopfront the overall dimension may be 8.00 m wide.

  The rod would show full size plan sections of the jambs at both ends but the glass would be two parallel lines therefore breaks in the straight runs would be made. For example, the left jamb plus glass would measure 200 mm then a break sign with 'add 3.00 m' then more glass and the left door mullion. Both the door stiles and glazing would also be reduced with a break in a similar manner and continue with the right mullion, glazing and break lines, glazing and right jamb. The shopfront rod would measure 1.00 m actual size and 7.00 m of additions noted, totalling 8.00 m in all. The same idea would be used for the counters and

Sheet 1 of 6			CUTTING SHEET			Job No. 0053
Item	quant	length	width	thickness	material	description
1	1	2000	200	18mm	MDF	plinth *tls*
2	1	898	200	18mm	MDF	plinth *tls*
3	1	2000	65	18mm	MDF	Top rail *tis.lll*
4	1	898	65	18mm	MDF	Top rail *tis.lll*
5	3	1218	900	18mm	MDF	Pot boards *lll*
6	3	900	875	18mm	MDF	Shelves *lll*
7	2	900	900	18mm	MDF	Ends *lll*
8	2	900	900	18mm	MDF	Partitions *lll*
9	3	682	900	18mm	MDF	Partitions *lls*
10	3	844	597	18mm	MDF	Doors *l2lls*
11	1	2862	200	1.8mm	Formica Black N128	Plinth
12	2	900	900	1.8mm	Formica Slate N147	Ends

Figure 9.5 Cutting sheet.

any other large item. Smaller items or those with many intricate cuttings would be shown in total full size without breaks.

As a junior setter out small parts of the project would be given out to be drawn up actual size on very long rolls of paper termed rods. For example the counter that goes the full width of the salon had to be drawn full size so that all the parts could be taken off for a cutting sheet. The cutting sheet would be quantified without any prices (see Figure 9.5). Information for the marker out is coded under the descriptions column. For example, the plinth is too long to cut out of one sheet of MDF board therefore two pieces will be joined together. The joint will be loose tongued, meaning that a thin piece of plywood (the tongue) 6 mm thick will be slotted into a groove in the end of each plinth. The code would be 't1s' on each of the plinth boards, meaning 'tongue one short side' (see Figure 9.6). MDF is not a durable material therefore it should be edged with another material such as hardwood. There are two types of edging: lippings, where the material is tongued onto the edge (see Figure 9.7), and edging, which is stuck square on the edge with or without pinning. The setter out writes all the information on the rod including the screw types and sizes and the code numbers of any fittings and the number of the component as it appears on the cutting sheets. Therefore the plinth is items 1 and 2 and on the rod will have '1' on the longer length and '2' on the shorter length. Both numbers will be in circles to identify the item.

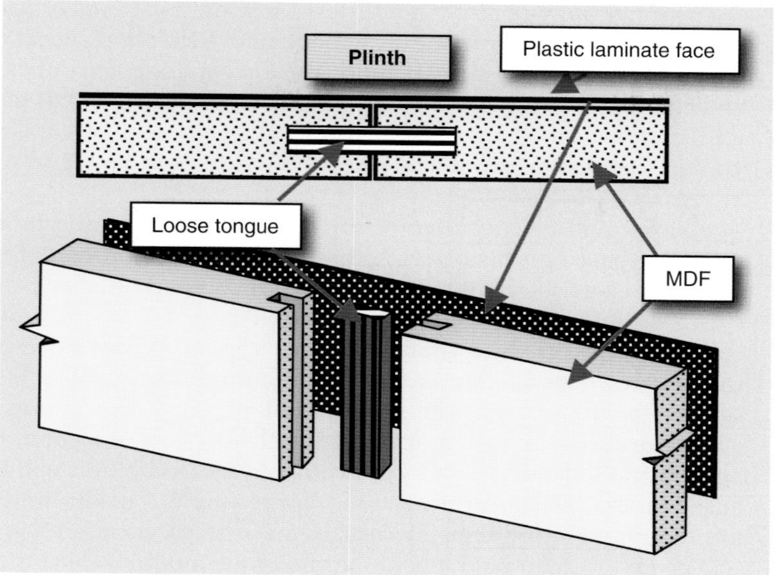

Figure 9.6 Plinth details.

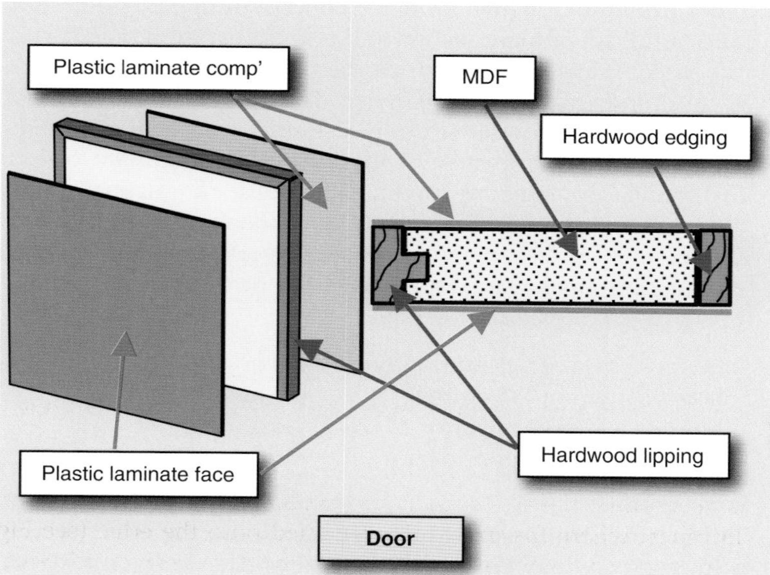

Figure 9.7 Lipping and edging details.

The setter out will also place orders for bought in items or special materials. It is usual to have one order book with each order sheet numbered and a self-duplicating sheet behind. The top copy is taken out of the book and either sent to the supplier or to the buying department. If it is the latter then it is termed a requisition order. Basically requisition means that you are requesting an official order be written out and sent off by the buyers.

- The buyer will be consulted as to which suppliers would be most appropriate based on history (whether they have been used before; it is a pointless exercise obtaining prices from a company that is perhaps the cheapest but does not supply on time). Another factor is whether the company has a credit account with the supplier. Where a company regularly buys materials or services from another company it is usual to set-up a credit account. The supplier will require confirmation as to the creditworthiness of the company using the services of a credit reference agency and the bank used by the company. A credit limit will be issued based on two months' transactions. For example, a credit limit of £5000 per month will need to be £10 000. Items or services ordered in month 1 will be invoiced for payment at the end of the month. Whilst the account is in action, the company has not settled/paid the supplier will still want to order materials in month 2. Suppliers usually offer a small percentage of the total for prompt payment, commonly 2.5%.
- The setter out will need to carry out a full measured survey (see Cooke (2007), Chapter 3). The setter out will be given the specification for the work from the designer. From the concept design the setter out will produce a full set of plan, section and elevational drawings known as rods. When complete, the setter out then produces cutting sheets showing each component, the type of material, the thickness and number of items (see Figure 9.5). The setter out is also responsible for ordering materials and fittings either directly with the suppliers or manufacturers or by requisition via the company buyer. All paperwork must have a job number on it to identify which costing account the materials will be set against (see Figure 9.8). Where the setter out is working on more than one job at a time a job sheet or time sheet has to be completed. For example:

2 hours setting out cash counter	job no. 215
3 hours setting out shopfront	job no. 210
1 hour ordering hardware	job no. 214
2 hours setting out gondolas	job no. 213

The time sheets would be given to the costing clerk to be associated with each contract so that at the end of the contract the hours can be monitored and the total cost calculated. Shopfitting companies do not normally charge by the hour, contracts are based on a total sum. To estimate future contracts the costing clerks gather the total hours taken on a contract which can be itemised. The cost of the materials can be calculated and the number of hours the setter out had taken.

Order Requisition

Order Number: **00137**
Job Number: **0053**
Date: 27th July 2009

To:
Comyn Chong Ltd.
63 – 67 London Road
London E7 2PQ

Goodwin Shopfitters Ltd.
[Works and Offices]
Unit 12 Plessey Way Industrial Estate
Hockley
Essex SS3 4PQ
Tel: 01702 639963 - 7

Please supply and deliver to our works the following materials on or before 20th August 2009:	Total £
6 No. bow handles [cat no. 06675] BMA finish @ £4.65 each	27.90
6 pr. 75mm x 18mm aluminium butts BMA finish @ £3.95/pr	23.70
1 box (100 No.) 25mm No.6 c/s screws aluminium BMA finish @ £5.95	5.95
Contact: Robert Cooke	
Value added tax @ 15%	£8.63
Total sum	£ 66.18

Figure 9.8 An Order Requisition.

The marker out is the person who reads the setter out drawings and cutting sheets and writes on or marks out the component parts ready for the mill to cut the materials to size. There may be several marking out stages; rough cut to shape first, then second cut and work on such as grooves and cut outs.

When all of the components have been cut, a bench joiner then assembles and makes the components fit together. Surface finishes such as plastic laminates will be glued on and trimmed by the bench joiner ready for delivery. The items will then be delivered to site where the site fixers position and cut in where applicable. All of the stages of work will be costed out and noted. For example, a cash counter of given dimensions and finish may have taken 60 hours labour and £280 of materials. The labour would be costed out as so much for the tradesman plus overheads, say, £20 per hour. 60 hours at £20 = £1200 labour plus materials £280 = £1480. A profit will be required to keep the company in a healthy financial condition therefore say 20% profit margin = £296. The total cost of the cash counter will be £1776. Future cash counters will be costed at, say, £1800 with a slight adjustment for size.

It is not practical or cost-effective to estimate shopfitting components on a 'take off' basis. The estimate will be based on historic work with an adjustment for inflation and the amount of work the company has: more keenly priced when work is difficult to obtain and higher prices when the company has enough work.

- The designer has provided an artist's impression of what the project will look like when it is complete. There will be no dimensions as such other than the overall ones taken at the initial meeting. Specifications will have been written as to the type of materials to be used and where applicable the manufactures names. The designer may also apply for Planning Permission where applicable such as for the shop signage.

  The Local Authority Planning Department will deal with Planning Applications for:
  - change of use – the premises in this example have previously been used for commercial purposes therefore no application required
  - Class 4: Illuminated signs – there is a general permission that a sign may be displayed either internally illuminated letters or characters on an unilluminated background or letters by 'halo' illumination. The signs must not:
    - have any intermittent light source, moving feature, animation or exposed cold cathode tubing
    - have more than one such fascia panel and one projecting at right angles.

  Further useful reading includes the free 'Outdoor advertisements and signs: a guide for advertisers' from the Government website: www.communities.gov.uk/documents/planningandbuilding/pdf/326679.pdf

Many of the Planning Regulations have changed in 2007 therefore it is important to use the most up to date information (see also Chapter 7).

When all of the materials have been ordered, assembled and or cut to size, and the fixers have completed their work, the decorators and finishing trades will start. The finishing trades will have been programmed to complete their parts of the contract at specific stages. For example, the main fixings such as the new shopfront, enclosure frames, falseworks, first and second fixings will be completed. (First fixings will be studwork, partition walls, electrical cabling, hot and cold water pipes, fire alarm conduits etc. Second fixings will be plastering, architraves, skirtings, electrical fittings etc.) Carpets or other floor finishes can be laid and covered with polythene sheeting to protect from follow on trades or programmed as the final finishes before handover.

As an incentive to complete the fit out either before time or on time the employer may offer additional bonuses on top of the normal wages. Working in the north of England meant the fixers having to work away from home. 'Digs' as they are known can be working men's hostels or budget bed and breakfast accommodation. To complete a fit out 'ghosters' are one solution. Although not technically legal for a tradesman to work that many hours straight through, it was a common solution to get a job finished more quickly by working all day and night. By continual working throughout the night the job may be completed days or even a week ahead of schedule, thus saving the company on accommodation costs, travel costs and allowing the fixers to start another job.

On John's hairdressing salon the shopfitters were local, therefore fixer's accommodation was not an issue. Sunday working, however, enabled the fit out to be completed more quickly. On a similar basis to save accommodation costs, Sunday working required extra pay per hour plus an early completion bonus.

When the shop fit out was complete a 'sweep through' was performed where the whole shop is cleaned and all protective polythene removed from the finishes. The contracts manager arranged to meet the client (John) on site to hand over the keys. John had to ensure the premises were covered by insurance as at the point of hand over the shopfitter was no longer responsible.

Chapter 10

A prestigious commercial development

The final chapter is a brief account/case study of a prestigious commercial development. Today, perhaps more than ever before, where companies are trying to maximise their profit lines there are still people who can see that educating the next generation is vitally important – see acknowledgements.

Ropemaker Place is an excellent example of how team working can achieve an outstanding building. Starting with a brief history of the area of London, the inception for the project, and the key team members, you will follow through with an account of the design and production stages. Much of the content of the book can be seen in context of a whole scheme.

The client has a very strong opinion regarding looking after the natural environment and sustainable development which is reflected in the design concepts produced by Arup Associates. Commentary has been given throughout the whole project with explanations where required of the construction techniques, technical and scientific issues used. Production issues including planning, compliance with Regulations, contract issues and all points between will provide a unique insight into this 21st century development.

10.1 Why demolish old buildings and erect another building in its place?

The built environment has developed over thousands of years. If every structure had to be saved and preserved, what a mess the country would be in! Disasters historically have enabled major new developments, but no one is recommending them as a development plan. Development will by definition be upsetting for some people. As we grow older we like to see familiar sights from our past.

It might be the row of individual shops, the butchers, newsagents, bakers, fishmongers, hardware stores and so on, all with their flats above where the shopkeepers would live. Ideal when they were built, but in today's society needs have changed. Generally people want cheaper prices, they do not want to spend lots of their time queuing in each individual shop, they want to be able to browse around gathering many different items and paying one bill at the checkout, therefore many smaller shops have closed down.

Companies want to group larger numbers of employees in one area, commonly in open plan offices that pro rata are less expensive to heat and cool and provide more space for the area of land they cover. Today's businesses require technology, kilometres of fibre optics and cables, air-conditioning ducts, booster motors and cooling plant, making it almost impossible to upgrade old buildings economically.

Land is becoming even more precious. In London even 700 years ago building land was both costly and in high demand. Later in the 16th and 17th century timber-framed buildings commonly had up to five stories, some added on at later stages to provide more space on the same plot of land. Originally much of London was owned by various religious orders where monks and nuns farmed the land in and around the city. Historic events (Henry VIII) forced the monasteries to dissolve, placing the land in private holding of landlords. Even today the landlords control large parts of London and lease the property. The high cost of property in London requires substantial investment and investors require returns on their money. British Land is one such company, with an investment portfolio in excess of £13.5 billion. They became the property advisor to the client 'Dominion Corporate Trustees Limited and Dominion Trust Limited as trustees of Ropemaker Place Unit Trust'. (For the purposes of this chapter British Land Plc. has been used when referring to the client. British Land Company Plc. are the property advisors to the client and in essence made the decisions on behalf of the trust company.)

In this chapter we will look at the whole pathway from conception through to the handing over of a prestigious new office development in the heart of London.

10.2 Inception for Ropemaker

Why did they choose to demolish existing buildings and develop a new office as opposed to refurbishing? The original concept required the demolition of two 1980s office blocks, one of which had replaced a 1950s office development, thus having a useful life of just 30 years.

A German development company in partnership with the NLA (New London Architecture) bought both properties with a view to demolishing them and erecting one larger office block on the site for their client Helical Bar Plc. The concept design comprised 23 levels of differing floor plates, ranging from 1500 m^2 in the tower to 2600 m^2 and 3600 m^2 in the podium levels, giving an overall floor space of 46 500 m^2. The architect had produced a design that met the client's needs giving a clean monolithic glass and metal façade to high specification office accommodation. To create the impression of space on what is a relatively small island site of 1.2 acres the design included a glass atrium to run through the structure. To maximise floor space and simplify the services, plant rooms were placed on the roof.

The German development company made the decision not to proceed with the project and placed the land with Full Planning Permission out to auction in late 2005. British Land Plc. had very successfully completed several other prestigious office developments in the business hub of London such as the Broadgate Centre and 51 Lime Street (the European headquarters for The Willis Insurance Group). They carried out an in-house feasibility study and brought in the designers Arup Associates to discuss various possibilities before putting their bid in. The whole process had only taking a matter of weeks.

Having put forward the successful bid in early 2006 British Land Plc. formally approached Arup Associates. They had previously designed five major developments for them since the 1980s therefore already had a good working relationship.

Unlike other architectural practices that employ consultant structural and services engineers Arup Associates are virtually unique in that they comprise architects, structural engineers and services engineers, all in one studio, ensuring smooth design development. This benefits the client in that only one contract exists between the team of designers and the client and the team is constantly on hand in one studio.

The Mace Group offer a similar package for the construction team and on the Ropemaker project four specialist divisions within the group:

1. Project Management (Mace PM)
2. Principal Contractor (Mace Construction)
3. Costing Consultants (Mace Sense)
4. Organisation Services (Mace Logistics Management).

See Figure 10.1. Another team from the Mace Group were just finishing another British Land development at 51 Lime Street; the Willis Building on the site of the Old Lloyds building (the centre photograph on the cover of *Building in the 21st Century*).

There were other specialists that the client brought in on separate contracts:

- Robert Townsend, who specialise in both hard and soft landscaping
- Gordon Ingram, a specialist consultant for issues of Rights to Light
- Gerald Eve, Chartered Surveyors and Town Planning Specialists

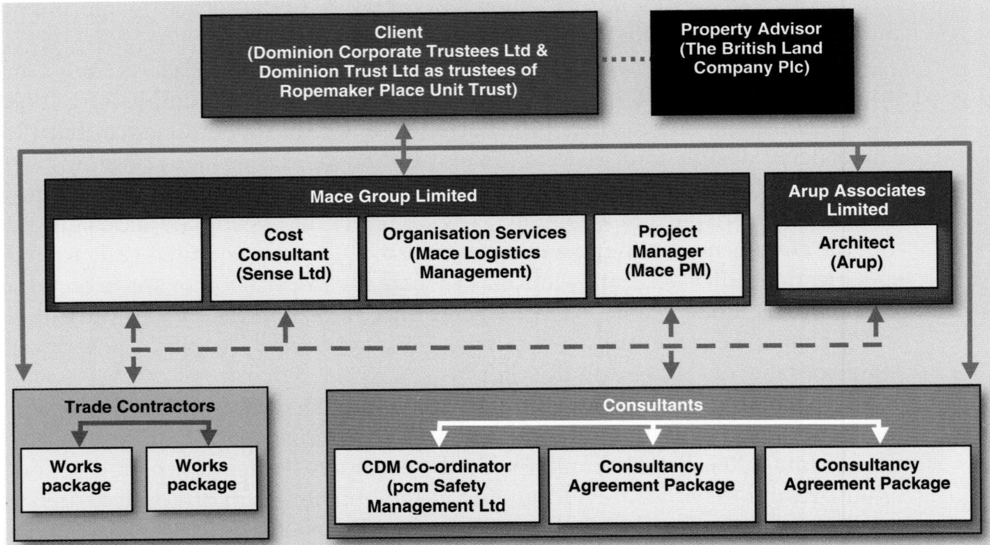

Figure 10.1 Client contractual relationship.

Plus several others including the Jason Bruges studio providing unique art-work on the walls of the atrium.

The client wanted a very high specification to ensure:

1 Quality of design – the building works matches the client's needs
2 Quality of build – is usually associated with higher costs, but quality is reflected in lower maintenance issues and a better working environment.

The next prime objective was a quick contract timescale (termed 'fast track') of about three years from start to completion in 2009. Although the site had Full Planning Permission based on an office development the new client wanted major changes. The designers had a two-stage proposal.

10.3 Stage 1: A second feasibility study

The first feasibility study had been prior to bidding for the site. This feasibility study involved discussion with the Planning Department of the London Borough of Islington and the London Mayor's Office. Working in London has particular constraints regarding City views termed as 'the significant views corridors'. The heights and shapes of the structures must not clutter the skyline therefore constraints imposed on the previous design regarding height were adopted. If the designers had rejected the constraints the whole project would have had to go through a lengthy Planning Procedure adding months or even years to the contract. The old adage 'time is money' is certainly true with speculative developments.

Town and Country Planning can be very complex and specialist consultants provide a service in a similar way to lawyers. By specialising in planning they acquire working knowledge of what Local Authorities will look for and accept based on historic cases. Planning is not hard and fast rules set in stone. The overall plan for an area will be decided by a Local Authority Town Planner. His or her preferences influence whether permission is granted or not. For example, an area such as Bath has a history of stone buildings built in local stone. However if the client wanted a green glass cube it is unlikely to achieve planning permission. There is an appeals procedure that enables a landowner to challenge such a decision based on the premise that as long as the structure complies with the relative regulations and is lawful (i.e. not a nuclear reactor in the middle of a residential area) the Local Authority would have to give good reasons for the rejection. Happily in this case the planning procedure although not usual went very well.

Planning Permission

The new client as previously mentioned wanted a very different design. The previous proposal had been based on leaving the existing basements and foundations and piling through to support the new superstructure. The concept of leaving the strip foundations and basement from the 1950s office plus a network of piles from the 1980s buildings did not meet the requirements of the new client or designers.

10.4 Stage 2: A new planning proposal

Arup returned to the London Borough of Islington planning department with a new planning proposal and further feasibility study including three new basement levels, thus significantly increasing the use of the land. The old basement and foundations would be removed and the new structure would have a raft foundation. Future developments would be able to use the raft system thus planning for the future. The new superstructure also met a radical change in design. A series of rectangular blocks integrated around concrete cores would enable larger floor plates for clients who require one floor trading and several smaller suites for the needs of other clients (see Figure 10.2). Arup Associates put forward the concept design for the formal approval by British Land Plc. including a visualisation which can be viewed on their website www.ropemakerlondon.com.

The three levels of basement floors were accepted by Islington Council and Partial Planning Permission was granted, enabling the contract to start on site. A major challenge taking about a year allowed the project to move forward whilst the detail of the superstructure could be finalised. This is a good example of Local Authority and a professional team working together. The Local Authority could have required the designers to submit the full design and go through the whole planning process before they could start.

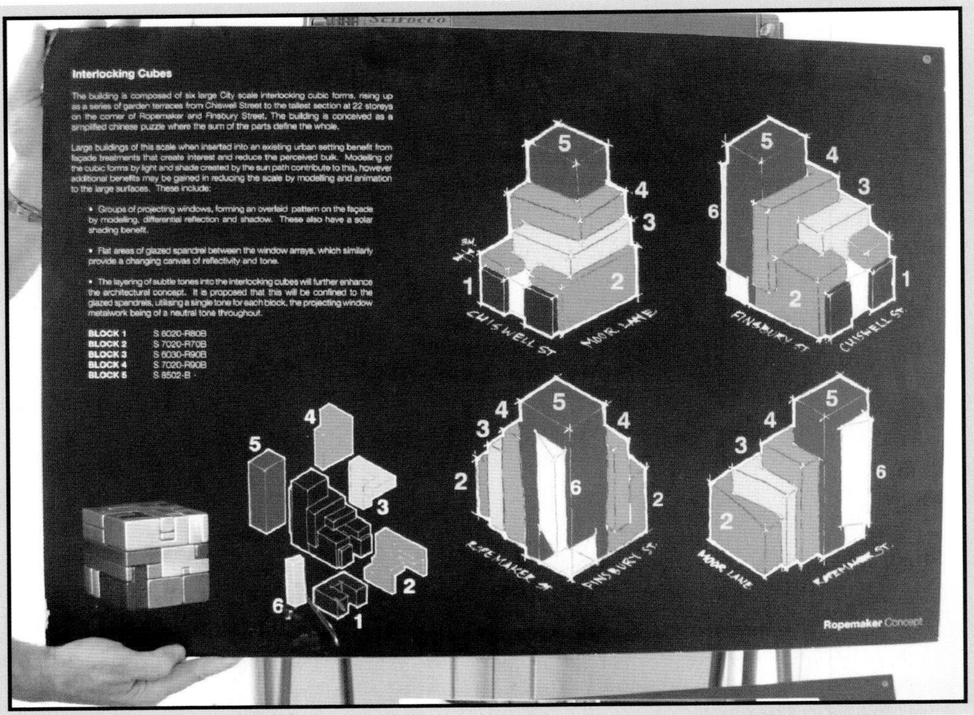

Figure 10.2 Concept design board.

So far we have:

- A client who has bought two existing office developments with a view to demolishing them and building a new office development for a specific blue chip company.
- An architect has designed the proposed new structure based on leaving the existing sub-structure and piling foundation through it.
- The client developer had taken a decision not to proceed and place the property including full planning permission at auction.
- British Land Plc put forward a successful bid and entered into contract with a design team, construction team, several specialist consultants and an array of trade contractors.
- Planning Permission had been granted albeit in two stages allowing work to commence on site whist the superstructure details were being finalised.

The new concept design

The island site by its very nature negates the need for a Party Wall surveyor. In effect the boundary comprised four roads under the possession of the London Borough of Islington.

Excavating basements required ground support. The existing gravity ground support was to be left in position and the engineers proposed secant piling on all four sides of the site. The principle is based on a series of augured concrete piles about 600 mm diameter reinforced with steel cages spaced at about 1.10 m centres. They are cast in slow-gain concrete enabling another set of piles to be augered in between them. The walls of piles are then capped with a continuous concrete beam tying them altogether at the top in effect a ring beam. – see Figure 10.3. By interlocking the piles, sands and other fines are held back, though depending on the substrate ground water may still seep through the wall. More detail is given under the contractors section. The raft foundation became the floor to the boiler room and much of the heavy plant enabling easier access and freeing up space on the roofs.

On a previous project, 51 Lime Street (the new Willis building), an interesting technique had proven very successful. Concrete cores containing the lifts and stairs would be cast using a slipform technique. They would provide rigidity to the structure and columns that the steel framework can be attached.

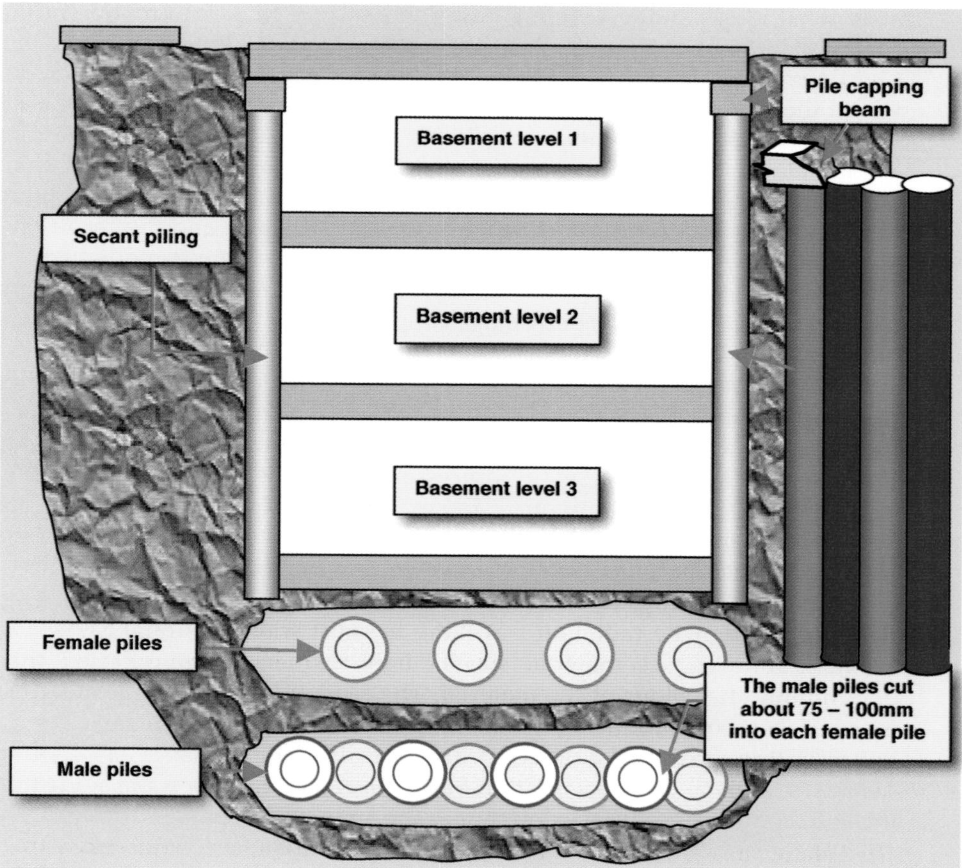

Figure 10.3 Secant piling.

The importance of light

Light is a very important issue and natural light is considered a right, hence the Right to Light Act 1959. As buildings become taller the problems become more complex. The origins of the Act date back to the Dickensian time of 1832 when buildings became so tightly packed together that the natural light through many windows became significantly reduced. At the time artificial lighting was mainly from oil lamps and gas lamps. Electric lighting only became available in the 1880s, but with bulbs costing the equivalent of £80 each in today's currency they were unlikely to be commonly used. The principles of the Act rest on an opening in a building having uninterrupted light for a period of 25 years or more. That does not mean that the owner can insist on the same amount of light to continue. He/she may have the amount of light reduced to a reasonable level. The word 'reasonable' comes up again, what is reasonable? Who decides, and how is it determined? That is why specialist consultants are employed. Every application is different and a matter of judgement.

The Building Research Establishment has produced a guide but that is exactly what it is, a guide. It is not law. There is also good information in the Metric Handbook and in the British Standard Code of Practice 8206-2:1992 Lighting for buildings. The yardstick approach is based on the part of a room where the opening allows light to enter. The measurement will be based on the surfaces receiving a minimum of 0.2% sky factor at a level of 0.85 m from the floor. The amount of light measured also takes into account light from any other opening that may not be affected by the proposals. Calculations are carried out based on the effect of the proposed structure and the result and compared as a percentage. Percentages are debatable in outcome though. A drop of 40% from 100 to 60% appears a greater problem than a 15% drop from 65 to 55.25% yet the lower percentage drop has ended up with about 5% less lighting.

Light also can be used to good effect in the external aesthetics of the building. Many new buildings have large glass façades from ground to roof. When the natural light levels are low the building contents is on display to the world. It is amazing how untidy many offices are when they are occupied. Some buildings use the mirroring effect to create the illusion of more space, but Arup Associates proposed a prismoidal effect creating facets similar to the way diamonds are cut. Double glazed window modules based on a 1.5 m wide unit size are angled to maximise the natural light entering the building. Sunlight is magnetic radiation travelling in short wave in straight lines. As the light bounces on a material, specific light frequencies are reflected, thus determining the colour of the material. The remainder of the light radiation is either absorbed, producing molecular movement and therefore heat energy, or passes through as in the case of clear or translucent materials. Glass used in windows tends to be relatively thin however depending upon the angle that the light contacts the surface determines the amount that passes through. Ideally the light should be at 90° to the surface. However, unless the glass is on the roof and at what time of year the optimum is rarely achieved. Daylight starts in the east and expires in the west passing through the southern plane.

On a 'green field' site it may be possible to orient the building to maximise the natural light and design fewer glazed areas on the northern aspect. The Ropemaker site has tall buildings a road width away on both the east and west elevations. The southern aspect has a wider outlook though.

Arup Associates' design incorporated tilted window units toward the north to reduce solar gain yet maintain the available sunlight and reflected light into the building. Solar heat gain is usually a problem with those working close to the windows on the south side of a building not only, but especially in the summer months. To reduce the solar gain yet allow the natural light to enter, the southern windows are tilted downward by 10° (see Figure 10.4).

To soften the façades colour was introduced to the design. Generally the use of unnatural colour can be fashionable but easily dates a building. For example, the spandrel panels of sheet asbestos or cement panels painted in blues and greens or the more durable painted and fired toughened glass used in curtain walling during the late 1960s and 70s. The client was understandably apprehensive. The design team had come up with an idea of introducing colour behind a gossamer of light-reflecting prisms (see Figure 10.5). Looking like glass marbles, the light is deflected through to the coloured lining on the inner skin. As the natural light moves around the building the surfaces subtly change from light to bands of colour. The tilted window units casting shadows further create constantly moving interest, something that is difficult to achieve with monolithic façade glazing. Light originates from the Sun and is reflected off surfaces, therefore by angling the glazing on the east and west elevation windows toward the north an increase of reflected 'borrowed' light can be achieved. The result is an increase in natural light and a saving in energy costs. This is just one aspect that has enabled an expected rating of 'Excellent' under the BREEAM assessment method.

The green issues

Modern buildings are required to meet the efficient use of energy. The Building Research Establishment (BRE) has produced a performance target for all new and existing buildings under the acronym BREEAM. (Building Research Establishment Environmental Assessment Method). Its adoption worldwide confirms the respect for the method. With a new build office projects a 15-page assessment document is used. Divided into a series of checklists, the estimator can compare the proposed project in contrast with the set standard. The Ropemaker project has been estimated in the best category 'Excellent' surpassing the standards.

The client's requirements are to be able to offer the highest specifications of accommodation whilst achieving the optimum in sustainable construction. With that remit the designer's in-house sustainability team were able to achieve:

- A projected 15% reduction in CO_2 emissions when compared with the requirements set out in the Building Regulations.

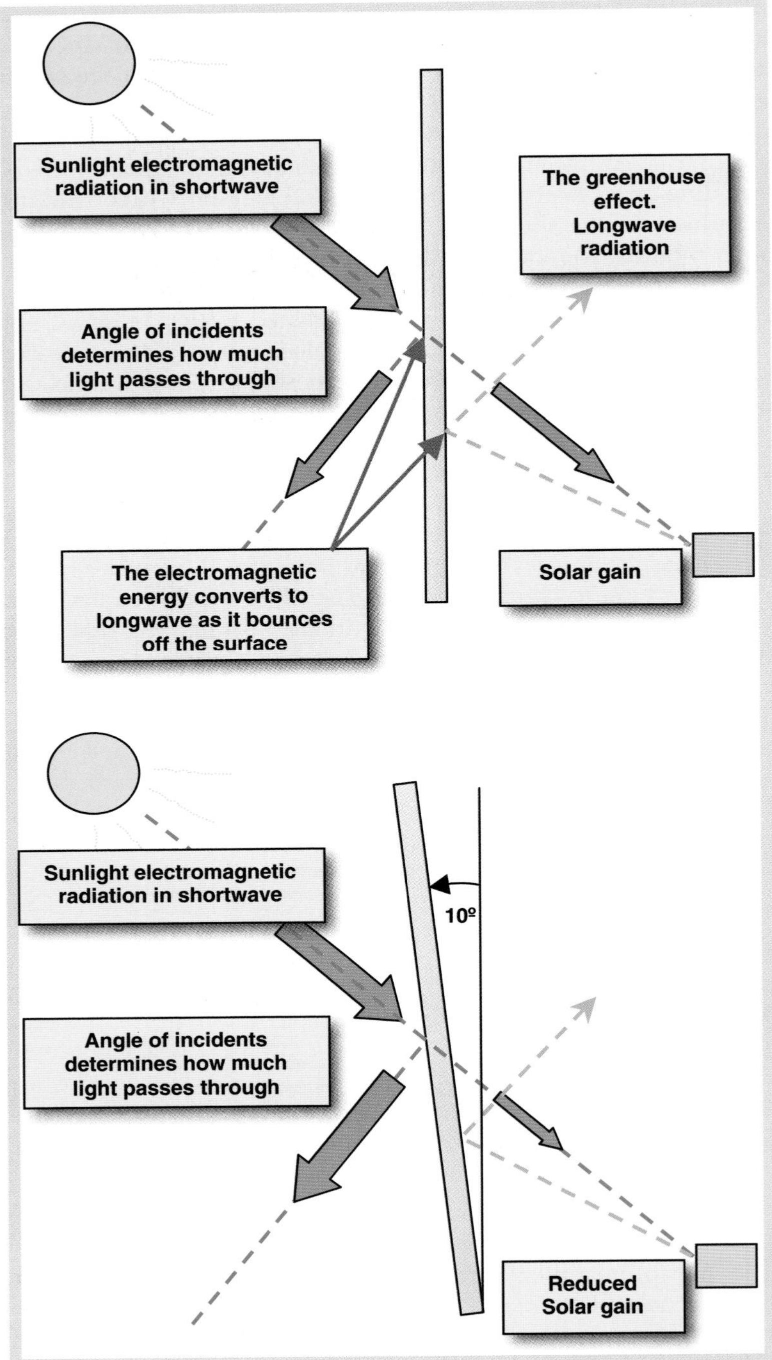

Figure 10.4 Solar gain v. natural light.

Figure 10.5 Light prisms.

- Collected rainwater for irrigation of the roof gardens and or for use flushing the WCs. It is estimated a water saving equivalent to 10,000 bathtubs per year will be achieved. As all piped water is cleaned ready for drinking (potable) quality the savings are both in water and the energy and chemical required in the cleaning process.
- Half of the roof areas will be presented as four roof gardens. Not just a bed of sedum, but a full spectrum of trees, bushes, flowers and grasses. As soft landscaping requires a different discipline to architecture and construction, a specialist consultant has the contract with the client. The advantages of a roof garden include aesthetic value to all that can see and use the gardens (see Figure 10.6). It is known that in a green environment people respond to the surroundings. That is one of the reasons why office buildings have large planters of lush green plants.

  The second reason is the cleaning qualities of natural plants and trees. Not just the photosynthesis taking place of the green leaves but the ability to capture airborne dust and trap it, forming earth. For example, look at how a seed of grass can grow with water and eventually trap enough dust to form a store of earth for its roots. You may have seen grass or even trees growing in roof gutters starting life from a seed contained in a bird dropping.

  The plant life will add to the biodiversity of the City. London is one of or possibly the greenest city in the world, with hundreds of parks and

Figure 10.6 Artists' impression of the completed structure.

gardens. By adding flowering plants to the rooftops insects and birds will have their natural environment in which to flourish.

Finally, green roofs hide the roof surfaces from the damaging ultra violet rays from the Sun. UVA and UVB over long periods harden polymerised materials such as roofing felts, neoprene and rubbers. Allowances for thermal expansion must be considered for sheet metals and masonry, the cause of many leaking flat roofs.

- Heating water for hand washing uses energy therefore a bank of flat bed solar water heaters have been placed on the highest roof deck. The solar heating is based on radiation therefore light. Although there is some advantage in July at midday, there can still be a useful solar input on a clear day in January. Clouds have the most effect, reducing the UV radiation more than the heated air. If you are several thousand metres up a mountain skiing, the temperature may be well below freezing but if the sky is clear the radiation will soon give you sunburn.
- Heating water by solar means also requires back up. For example, what happens during dull days or when snow covers the collectors? The building also has biofuel boilers to keep the water at the required temperature.

Working on thermostatic control, the well insulated hot water cylinders only use the minimum of energy. The biofuel boilers can be fuelled by either natural gas (one of the cleanest of fossil fuels) or pelleted wood. Commonplace in Germany and Austria, the technology is well proven. The pellets are available in the UK utilising saw dust from softwood timber mills. The dust is highly compressed and has to conform to European Standard CEN/TS 14961. The calorific value is 4.8kWh/kg therefore 3 tonnes/6 m^3 should produce about 14.4 MWh of renewable energy. The pellets are gravity-fed into the boiler on electronic demand from the hot water storage cylinders and the boiler thermostats.

- Next to the flat plate solar collectors is an array of 153 m^2 of photovoltaic cells. Typically an array will produce just over 3 A at 0.5 V per cell (100 mm × 100 mm) therefore a potential in excess of 7.6 kV at 3 A. Not enough to power the building, but certainly a contribution. Between the solar plate collectors, biofuel boilers and photovoltaic arrays, there is a saving of more than 20% in CO_2 emissions when set against the 2006 Building Regulations. Reducing energy demand not only reduces the CO_2 emissions but should extend the time left of the non-renewable fuels.
- Modern office buildings tend to be very well insulated and subject to greater cooling than heating demands. People generate heat, computers have become so power-hungry that they generate a lot of heat. (My own PC has a 500W power unit with 2 fans continually cooling the microchips.) With the amount of PCs, photocopiers and laser printers all giving off heat the problem soon mounts up. Having opening windows in City buildings is not ideal. The smoke from cigarettes, although by law it should be out of the buildings, tends to cling on upward air currents and permeate back inside. It is a well known phenomenon and in many States of the USA smokers must stand a minimum of 12ft (4 m) away from any building. (The smoke still blows in if the wind is in the wrong direction.) The air quality at street level in London is not good, especially during the hot spells of summer. To overcome the two issues of heat build-up and external fumes, the design incorporates air intakes from roof level, making use of the cooler air during the summer months.
- Further energy saving considerations include full showering facilities for those who wish to cycle to work. (Nothing is more off putting than sitting all day in sweaty clothing), and easy access to three forms of public transport. For those of the 3000-plus people living in the Barbican, it could be a very short walk to work.
- Where possible, all building materials used in the new structure will be recyclable. Even the demolition materials have been recycled and used in the new structure.

So far we have:

- The client has provided the designer with:
  - a comprehensive Office Design Brief
  - a Sustainability Brief.

- The designers have produced a concept design for the approval of the client. The client signs a document confirming the approval.
- The building surveyors have carried out an additional survey based on the removal of existing foundations and basements.
- The architects and structural engineers have produced a scheme based on:
  - three basement levels
  - slipform service towers
  - steel frame superstructure
  - specialist curtain walling
  - tilted double glazed window units with coloured inner sheets for spandrel panels and non-vision areas
  - rigid high performance foam insulation achieving excellent U values to external walls.
- The scheme is very energy-efficient and is expected to achieve the BREEAM 'excellent' rating based on the use of solar energy technology, green roofs and innovative design.

The contracts

The contracts used on Ropemaker comprise bespoke Management Agreement (MA), not to be confused with a Client Management Agreement (CMA). The client's legal department modified a CMA to include the extended services:

- Project manager (Mace Project Management)

  As the title suggests, the team, led by the project director Jonathan Foster, had the overall responsibility for ensuring the project was completed on time. Based in offices close by the site enabled their contractual duties as project managers and construction advisors to the client to be carried out. In effect the hub or kingpin of the construction programme.
- Construction manager (Mace Construction Managemen)

  The management team based on site orchestrates the army of trade contractors who actually construct the building. Essentially the construction management are responsible for organising all of the trade contractors by careful planning including all aspects of health and safety on site. The team of construction managers ensure the work is carried out at the correct time and of the correct quality. A major part of construction is to work safely therefore the team would collect method statements from the trade contractors and suppliers before any works are carried out. That includes the issuing of 'hot work permits', 'lifting permits' etc.
- Cost consultant (Sense Ltd)

  Another team based with the project management in offices close by the site. Their function is that of the PQS. Almost as the chancellor is to Government, the cost consultant advises the client on the payment of works completed. Unlike standard building contracts where there are PQSs working for the client and CQSs working for the contractor, MA contracts only require a PQS as each trade contractor is responsible for invoicing the client. The fees for the management teams are pre-agreed.

- Employer's agent and fixed price contractor for all site establishment and logistics (Mace Logistics Management)

  Working closely with the construction management team they ensure the shared plant such as tower cranes, hoists, lifting equipment etc, accommodation, canteen facilities, first aid room and so on are available.

As you can see, the lines of direction are very different from, say, a JCT Standard Building Contract (SBC). The client has been in existence for over 150 years although a merger with Union Property Holdings in 1970 marked the start of the development company of today. Almost 40 years of construction-related development has meant the company has a strong in-house team able to provide a detailed client's design brief. This is in contrast with the clients, who require a one-off building with little to no experience of the complexities of architecture or construction.

The MA enables the client to keep full control of both the design team and the construction team with more of a 'hands on approach'. The client needs to be experienced as he/she will be in overall control. In contrast the JCT SBC is designed to enable the client to have a single contract that the lead consultant acts and controls on the client's behalf. This is more suited to a client who has little to no previous experience.

On the Ropemaker project the client had invited the designer based on experience of several previous successful office designs. The client/designer rapport had been well tested and enabled confidence from the outset. This is very important when selecting the designer. An analogy was given by Paul Dickenson, director of Arup Associates: the customer has asked for a high specification TV and thought he was getting the high quality item on display. However, when it was delivered, the TV was a cheap copy with the same specification. The same could be said of building design. The client has asked for a high specification but will he/she end up with the quality? British Land Plc know what they want and are confident the designer will provide it.

The design practice has regular team meetings, ensuring that all members are constantly working on the latest details. Having the main disciplines in one studio has a significant time advantage over teams based in various buildings and perhaps great distances apart. There are three architects and one engineer from the design team based on site to deal with day-to-day issues. In addition to the team meetings there are weekly site meetings with the principal contractor. The meetings provide feedback, enabling adjustments to be made where necessary. In contrast with the JCT SBC where the designer issues a variation order (VO) with the CMA, the designer issues a 'change request'. Notice the subtle difference in approach from 'order' to 'request'. The end result will hopefully be the same but the approach is very different. This also applies to the architects instructions (AI). This now becomes a Construction Management Instruction (CMI). Where a JCT SBC has been used, the architect is the main person working as agent to the client. This is in contrast with the CMA, where the architect is a team member who advises the client of the best method of approach.

As the project proceeds, the work is reviewed on a monthly basis by:

- the client
- the Contract (design) Management co-ordinator
- the Construction Management
- the Designer

and valuations are prepared. Valuations are based on the completed work on site plus any significant high value items that may be in storage for protection.

To ensure the client only pays for the completed work as per the contract documents, Mace-Sense measure the work and compare it with the trade contractor's claims. The information is then passed to Arup Associates to verify and confirm it complies with the contract documents. The confirmation is then passed to the client, who authorises payment (see Figure 10.7). Note that the client has the contract with the trade contractors, therefore all payments would also be direct. Each trade contractor is responsible for their own part of the contract, including obtaining and delivery of the materials, labour and plant. They are also responsible for the removal of packaging, waste and any materials not used. The Construction Management ensure that each trade contractor has shared facilities and storage space on site. This is something that will be discussed in more detail in section 10.8, Site set-up.

When the contract is complete the Construction Management Project Executive on behalf of the principal contractor and the director of the design practice will jointly sign off the building. This is in contrast with a JCT SBC where the designer (commonly the architect or lead consultant) would issue a Practical

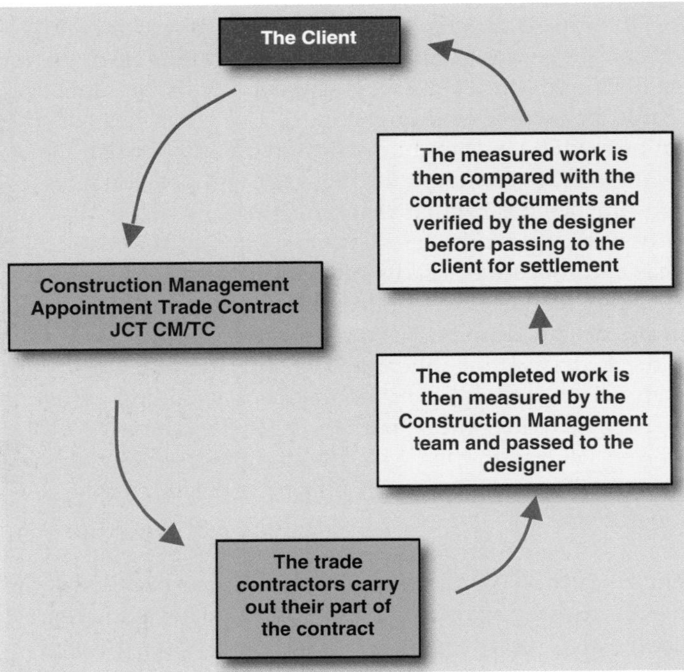

Figure 10.7 The financing cycle.

Completion Certificate. On this project the two main parties will confirm that all of the work has been carried out correctly and in compliance with the contract documents. This is a significant date as it then starts the period of defects liability.

The client can now take possession of the work.

The principal contractor and the designer will carry out a schedule of defects at the end of the defects liability period as stated in the contract, or within 14 days of the expiry of that date. The various trade contractors will be able to make good (which means to repair or replace the defective materials or work). If the trade contractor fails to make good, the retention moneys will be used to pay another trade contractor to make good the work. Any extra costs will be recovered from the trade contractor who carried out the original work. In practice, depending upon the amount of work available generally and the amount involved, some trade contractors ignore requests to make good their work as it means they will have to stop new work in order to carry out the remedial work. That can cost more in penalty clauses than the loss of the retention money. In Construction Management Contracts the defects are the responsibility of the Trade Contractor under their own contract. The principal contractor only has a duty to advise the client of such defects. After the contract liability period is complete and the handover of possession has been given to the client, the client will then have to provide a licence for the Trade Contractor to remedy the work. The official requirement is to satisfy the insurance cover for such remedial work.

Defects are generally picked up/noted during the construction period. On the Ropemaker Project the senior construction manager walks around the whole site on a daily basis checking on health and safety issues and any noticeable damage or defects. In addition there are several construction managers who are responsible for their particular part of the project. With so many checks being carried out there is no requirement for a Clerk of Works. We will be looking at the issues of site management later under section 10.8, Site set-up.

As a general point when things go sour and end up in court it can be very costly in both time and money. The legal fees can run into hundreds of thousands of pounds, even millions, therefore arbitration is usually more acceptable. Contracts will normally have who will arbitrate and where written into the contract. Both parties will sign to agree to arbitration, but in English Law a party may still sue for damages even after arbitration.

Some cases go through the courts. According to Daniel Atkinson Ltd., who are arbitrators, adjudicators and mediation specialists, in the case *Westminster City Council* v. *Jarvis & Sons Ltd* (1970) the defects had been discovered and the contractor had remedied them during the defects liability period. The lead consultant then is required to issue a Final Certificate. The contract may state that the Final Certificate is a termination point for the rights of abatement. The Final Certificate should not be signed whilst there are ongoing issues of defects. There are exceptions:

- If the contractor acknowledges the defect in writing, as long as they are not trivial.

- That the lead consultant is satisfied that the works are reasonably complete. (What is reasonable though?)
- If there is enough retention money to remedy any known defects during the set period.

In all cases the above should be carried out only with the approval of the client as the certificate is a legal document and as such has consequences.

In addition, during the above case, according to Stephen Bickford-Smith and Lana Wood (in Speaight, 2004), the term 'practical completion' is not defined in the contract. They continue; it has been said by Lord Dilhorne in *Westminster City Council* v. *Jarvis & Sons Ltd* (1970) that it does not mean the stage when the work 'was almost but not entirely finished' but 'the completion of all the construction work that has to be done'. The case went to the Court of Appeal, where Salmon LJ ruled that if he took the word 'completion' to mean every last detail then it would become a penalty clause and thus unenforceable. As you can see, contract law is a specialised subject and where possible it should be left to legal minds to write the contracts – and use arbitration and not the courts of law.

The client has a right in common law to raise a defence of abatement against a Contractor requiring full payment for works done. (There are exceptions, but they must be written in the original contract as an exclusion clause and have the signed agreement of both parties.) *Slater* v. *Dumequin Ltd* (1992). Abatement means to make less strong or put off (for example the noise abatement society want to reduce unwanted noise) and only applies to defects which are patent (obvious) at the time that the full payment is due. For example, the client takes possession and finds that a tap in the cloakroom continually sticks open. The plumbing contractor should have had it brought to his attention during the snagging (where the main contractor or principal contractor tours the works with the client, inspecting the completed work and compiling the schedule of defects prior to signing off for handover). The plumbing contractor cannot demand full payment until he/she has the remedied the defective tap during the defects period. The defects period commences at the date of practical completion for a period of time as stated in the contract. It is usually six months, but another period may be set in the contract.

Moneys are withheld, termed 'retention' by the client in case there are any defects found at a later date. The amount of money is normally a percentage of the total amount payable for the work carried out and written in as part of the contract. However, to what extent can retention be made? For example, what would happen if, say, the supplier of the concrete for the secant piling had not supplied the correct grade of concrete? Highly unlikely but possible. That is why concrete samples in cube form are tested at 7 days and 28 day after the concrete has been placed. If there is found to be a defect in the quality it would normally be noticed before further work is carried out. What, though, if the roofing contractor has not adequately sealed the roof and used insufficient adhesive to bond the materials? A subsequent leak, say over the Christmas holiday shutdown, may penetrate the building, damaging ceilings, perhaps office equipment, the carpet and sub-floor equipment. To what extent does the defect liability extend? Just the cost of repair or the cost of accessing the defect and the cost of the other damage including the specialist to carry out the work termed 'consequential loss'?

Claims against defective materials or workmanship must be placed in writing within 28 days of the defect being known. The claim should be clear as to its nature and the remedy being sought. For example, is the claim financial and acceptance of the defect? An analogy could be you bought a pair of new shoes and after wearing them for one evening it became apparent that the shoes had not been cut from the same piece of leather as the grain did not match. You like the shoes and could put up with the mismatch of leather, but you want compensation as the shoes are not of the required/implied quality. Likewise a defect in a building material or quality of workmanship may be accepted as it could cause more damage rectifying the problem than leaving it in place. A monetary resolution may be sought via discussion and agreement.

In standard JCT contracts after the defects liability period has expired and not later than 14 days from that date a schedule of defects should be produced by the main contractor. The sub-contractors then have a reasonable time to rectify or remedy the issues. When the defects have been resolved the lead consultant can then issue a 'Certificate of Completion of Making Good Defects'. The main contractor can then request the remaining retention monies to be released.

In the case of the Ropemaker project they are not using a standard JCT contract.

The client British Land Plc. is a very well established developer, continually involved with major construction projects. They have produced, with legal guidance, their own bespoke set of contracts based on their own needs and commercial practices.

Before proceeding with any work on the project two contracts were signed: Construction Management Agreement (CMA) and Client and Construction Manager (C/CM) contracts. This created a legal contract between:

- British Land Plc. (the client) and Arup Associates (the designer)
- British Land Plc. (the client) and Mace Ltd (the construction manager).

Other parties are written into the contract, including:

- the Construction Design and Management co-ordinator who will ensure the project conforms to the CDM Regulations
- the Planning Supervisor
- the Principal Contractor.

The client will then enter into contract with the various contractors who will be employed. They are termed Trade Contractors, who will be bound by Trade Contracts. Their contract document is the Construction Management Appointment (CM/A) where the client has either:

1 authorised the Construction Manager to act on his or her behalf
or
2 has a direct and separate contract with the trades, therefore using the Construction Management Trade Contract (CM/TC).

As with all contracts there are clauses stating what should happen if say someone dies or for some reason has to terminate the contract.

The contract clearly defines the intentions of the identified parties and the documentation that makes up the contract comprising:

- the Recitals
- the Articles of Agreement – these include:
  - the obligations of the Construction Manager
  - the sum of money that the client will have to pay out to the Construction Manager
  - what should happen if the Planning Supervisor should die or change his or her job.
  - In the 'Articles' it states that the Principal Contractor under the CDM Regulations will be the Construction Manager. If for some reason the Construction Manager cannot continue the client may then appoint an alternative Principal Contractor to satisfy the requirements of the CDM Regulations.
  - The final article sets out what the procedure will be if parties find themselves in dispute and who the adjudicator will be
- the Conditions
- The term 'Appendix or the Schedules to any clause' means that it should be read as a 'Clause of the conditions'. For example the Defects Liability clause will be found in the Appendix.

An additional clause states that the defects liability period will commence for any trade contractor on the date of the Interim Project Completion document. When the document has been signed the client is then responsible for insuring the work. The document also identifies the ending of the trade contracts known as the construction period.

As the Ropemaker contract is based on a CMA method, when the project is complete the principal contractor and the design consultant will confirm in writing to the client that the work is:

- of the contracted quality
- meets the specification set out in the design contract documents.

This is in contrast to a standard building contract, where the lead consultant issues a Final Completion Certificate.

The contract confirms that the documentation should be taken as a whole and not in isolation. Words or phrases have been given a definition, such as:

- the Agreement
- All Risks Insurance
- the Appendix
- Articles of Agreement.

For more information on contract phrases and terms it is suggested that you buy a blank copy of the various standard building contracts from the RIBA, ICE or RICS bookshops.

Standard building contracts contain a wide range of clauses identifying potential scenarios and the methods of approach to overcome them. For example, what should happen if the client wants to take partial possession before the Interim Project Completion Certificate (PCC) is issued to the client? As it will have a direct effect on the defects liability period, a clause sets out the relationship with the PCC and the Trade Contracts. As the various contracts are used any anomalies or issues that become apparent are sent back to the authors or legal departments for consideration. For example, any changes in legislation that will have an effect on a contract will become amendments or revisions therefore contracts are continually evolving.

As mentioned in Chapter 1, contracts are intended to simplify the intentions between two parties by writing the agreement down. However, even simple contracts can have so many variables that it is difficult or nigh on impossible to foresee all situations. Arbitration is preferable to taking full legal action, although there is ample evidence, given the number of legal practices specialising in construction law, that things still go wrong.

10.5 Costings

From the inception stage where British Land Plc were considering placing an open bid for the land, Mace Costing, now renamed as Mace-Sense, were involved. The first stage from inception is the feasibility stage where the client needs to know an approximate cost for the project to establish whether it is within the budget set aside for the proposed development.

Calculating the approximate costing

There are various methods:

- Cost per m^2 – the form of construction would be decided by the architect/designer and the structural engineer. The Professional Quantity Surveyor (PQS) will use historic final costs and calculate them based on the floor area. The calculated cost will be an approximate cost per square metre including plant, labour and materials, but not the land (see Chapter 5).
- A similar method would be used for cubic metres, where the height of the building would also be considered.
- Where there are numerous rooms of a similar type, such as an hotel, school or hospital costs per room may be used as an approximate costing.

In the case of Ropemaker, the approximate costings were based on the design of the original concept and used as a benchmark. Although quite old by 2006, as the project had been shelved for about two years, the PQS (Mace-Sense) reviewed them and compared them with the proposed quality of construction intended for the new design. Final costings on previous British Land projects suggested that the finishes required would cost more, so the cost per square metre was increased to reflect the new specification. The gross floor area and number of floors had changed from the original, therefore further adjustments were made.

What is floor area though? How is it measured? Take an analogy of, say, a leg of pork. You say to the butcher you want about 2 kg. That is an indication of the amount measured as a weight (or more correctly a mass). It is the easiest method, as using lineal, area or cubic dimensions would be very difficult. However, does the 2 kg include the bone, fat and skin or just the meat?

Buildings are structures, so do we measure floor area as the area covered by the structure? Do we include the thickness of the walls and finishes? Do we include the stairs as floor areas?

Returning to the examples, at the shops the fresh trout look good. The cost per kilogram is displayed in a price tag on the fish. You select two trout and the fishmonger places them on the scales. He tells you the cost and then asks whether you want the fish gutted, and the heads and tails removed? You end up paying for the guts, head and tail even though you do not want them (unless you have a cat). With the building costs, especially office blocks, what do you want as floor space? What will you actually pay for if you are renting/leasing the building? One solution is to use the RICS Code of Measuring Practice. However, things can still go wrong. Eleanor Richardson, a solicitor of KSB Law LLP, cites the case of *Kilmartin SCI (Hulton House) Limited* v. *Safeway Stores plc* [2006]. Safeway had agreed a minimum internal area and during the works requested a lift be provided. No allowance for the lift area was made therefore they were going to pay for the area the lift took up as internal area. In the legal case that followed it was decided that the premises had been measured in accordance to the Code net area and lifts were correctly excluded from the internal area. Safeway Stores considered they had not received the amount of space in the agreement and could have possibly withdrawn from the contract leaving the developer a loss of £1m development cost and a loss of £600,000 per annum floor rental. The difference the lift made according to the judge was 3 m^2.

It is important to agree in writing exactly how the floor area is to be measured. Will it be gross floor area or net area? Will the office floor be partitioned off therefore will the thickness of the partition be included as floor areas? Will the floor areas be measured to the RICS Code of Measuring Practice?

Generally when a client rents office space or retail space the toilets, stairs and lifts are included in the lease but excluded from the internal area. For example, an office block in London could cost between £60–75 per square foot (I know we have been in metric since the early 1970s, however it sounds less and most companies charge at that rate). That means a wastepaper bin could cost about £70 per year in floor rent. Prestigious buildings, however, are charged at the per person rate. In the City of London in 2008 £1300 per person per annum is a typical rate. That would include services such as toilet areas, lifts, stairs, escalators etc., though specialist client requirements such as air conditioning would be an additional charge in some contracts. It is the letting costs that influence the building costs. For example, a prestigious building such as Ropemaker will be leased out to blue chip clients who require prestigious offices with excellent environmental credit in the heart of the business area of London.

10.6 Pricing the contract: 'taking off' and 'bills of quantities'

On a project of the size and complexity of Ropemaker it would be infeasible to design the whole project through to production drawings before handing it over to the PQS for a formal take off. Unlike traditional housing projects where the designs and materials do not significantly change, production drawings, also known as detailed drawings or working drawings, can be produced in a short time period based on similar projects. With today's computer aided drawing techniques, the detail blocks can be used and adjusted for dimensions. Window and joinery manufacturers together with plumbing, drainage and sanitation manufacturers provide software blocks that can be brought in ready finished, again reducing the time required to produce detailed drawing significantly. For more detail on Bills of Quantities see Chapter 5.

Back to Ropemaker. The client has an approximate costing for the proposed work based on costings prepared for the previous design plus a percentage increase to allow for the superior quality of finishes required. British Land had specific suppliers in mind such as Swift Horsman who they had used on previous developments. They produce modular pod units for the wash room areas. The unit design incorporates hand basins pre-fixed into polished stone counter tops with all the taps, piping, wastes etc. fitted and ready for connection in the room. The pods are delivered wrapped and protected for transportation and costed per pod, saving time, specification and drawings when compared with traditional contractual work. The quality is also assured as they are a factory finished item.

Mace-Sense (PQS) carry out three cost planning stages:

1 Preliminary estimating stage
2 Concept stage
3 Scheme design stage where the client makes the final decision

Preliminary estimating stage

Preliminary estimating included approximate costings as previously mentioned. In addition, unit costs for major plant and materials such as concrete would be gathered. On this particular project the crane contract was direct with the client, but unless the main or principal contractor intends to use in-house cranes there would normally be a specialist sub-contract via the principal contractor. There are issues though when using a CMA type of contract. As such, the management team are contracted to oversee the construction of the project. The construction is carried out by an army of trade contractors who for the most part will supply and fit materials as per an individual contract with the client. Items of large plant such as cranes, scaffolding, hoists etc. are shared by any trade contractor who requires it, so the principal contractor would provide the items.

A plan or 'method statement' from each trade contractor is required to enable the principal contractor to plan the requirements of the trade contractors throughout the project. However, it is impractical to wait until all of the method statements have been prepared, so the planners have to make educated guesses as to the plant requirements and the duration the plant is required. To provide a costing for the plant and site accommodation, right down to how much the telephone calls are expected to cost, would be impossible, so a percentage of the total estimated cost for the whole project is provided as a total budget. As with all budgets, allowances or contingencies must be made for unforeseeable events.

Cranes are by their very nature subject to periods where they cannot be used due to the weather. If the principal contractor has to budget for the cranes he will need to put in a sum to cover possible stand-down time due to wind, rain or snow – an inflated price. If the client has the contract with the crane hire company he will perhaps have to pay for the stand-down time, but if there is a low number of hours/days involved it can be cheaper than paying the extra contingency via the principal contractor. The crane operator/lifting specialist contractor produced a 'schedule of common lifts' comprising two columns headed 'load' and 'method' showing different types of lift and the methods to be adopted. For example, under 'load' – concrete or muck skips; 'method' – skip chains 8 tonne capacity. The schedule includes points or issues to be considered such as 'beware rusted bases on stillages and lorry type skips'. The lifting specialist also supplied a document containing photographs and greater detail of lifting procedures as part of the method statement. Where mobile cranes are used on site, permits are required outlining the details of each crane, the period of hire, and maintenance schedules including when and what must be tested and certified. Copies of the method statement, mobile crane permits and driver licences/certificates were kept by Mace, who acted as agent to the client. The contract for hire was between British Land Plc. (the client) and the crane and lifting specialist. Mace issued their own documentation such as 'crane lifting plans' and maintenance records as evidence that all safety and maintenance operations had taken place on a planned regular basis.

Concept stage

Costing at the concept stage is more straightforward. The designer will invite specialist suppliers and supply and fit contractors to tender for specific elements of the project, for example the curtain walling. The designer will select a system that would meet the client's requirements and invite the company representative in to discuss the project. If the technical aspect meets the criterion the designer has set the company would be asked to provide a costing for the work. The PQS would look at the approximate costing for that part of the contract and if it is within the budget confirm the information to the designer. At the concept stage detail is not required therefore the time element is kept to a minimum. The main parts or elements of the project would be tendered for and compared with the original approximate costings. As you can see, it is in stark contrast with 'taking off' and Bills of Quantities.

Scheme design stage

The third stage is the scheme design stage. The designer has assembled the concept design and selected a group of suppliers and fitters. Specific trades are at this stage not needed. As it is the client who will be paying for the project, the whole design is presented to the client for approval. Arup Associates produced computer-generated virtual reality programmes enabling British Land to electronically walk through the finished project. The detail is astonishing and can be viewed on their website: www.ropemakerlondon.com. In addition, sample boards and concept design portfolios enabled the designer and client to fully consider the whole project before any detailed drawings or construction details were produced.

So far we have looked at some of the paperwork side of pre-contract work and sub-structure work:

- Two main types of contract with the client:
  - Client Management Agreements (CMA)
  - Construction Management Trade Contracts (CM/TC).
- The design team.
- The project management team.
- A cost consultant.
- Approximate costings have been established against a bench marked costing with an allowance for the design quality changes.
- Feasibility, concept and scheme designs have been produced and agreed with the client.
- Specialist suppliers and fixers have provided tenders and they have been compared with the approximate costings.
- A budget Figure has been given to the principal contractor to cover plant and equipment plus accommodation and personnel.
- Detail design can now be produced and Full Town and Country Planning can be applied for the superstructure.
- An application for Full Plans (Building Regulation Approval) can also be made to include the superstructure.
- The sub-structure work is well advanced.

The next section will look at the paperwork side from the principal contractor's perspective.

After being awarded the contract as Construction Manager Mace Limited became the principal contractor. They assembled a team of experts under the project executive Jonathan Foster. Projects of the scale of Ropemaker need experienced people who specialise in their profession of management. With a contract time of less than three years from start to handover and large penalty clauses there is no room for experimentation.

The project was divided into stages:

1. pre-contract planning
2. post-contract planning.

10.7 Pre-contract planning

Before any work is carried out the site must be made secure. But how much money should be allocated for, say, the hoarding? You could measure the curtilage of the site and produce a materials list, then go out to tender for the materials and cost of labour – all a very lengthy and time-consuming process. Mace, however, use a very powerful spreadsheet template that enables fast and efficient analysis of the whole project and presents it as an overview. Strategic data is entered into specific cells and the programme then fragments into a series of sub-totals, enabling the cost controller to provide guide or target sums. For example, given just a few details including total sum, numbers of floors and footprint area, plus duration of the project etc. the software provides guide prices for the hoarding, plant, temporary welfare and site accommodation, materials, right down to how much the photocopying costs should be. I haven't been able to go into detail as the software has been developed in-house and gives the Mace Group a strong advantage when tendering for large contracts.

The cost controller for the project can then adjust schedules of rates and make strategic amendments specific to the project in hand. The spreadsheets provide a rolling template with a master worksheet, providing a constant overview with linked individual worksheets giving fine detail of the worked up prices. Fluctuations in labour and materials costs can be continually monitored providing weekly or monthly updates to the budgets. Working with the operations director a pre-production plan was formulated. As with any contract a site visit is essential. Information about the logistics of everything, including:

- Delivery of materials – are there any restrictions to the size and weight of vehicles going to and from the site?
- Where will the vehicles be off loaded?
- Where everything will go during the programme – one thing is a drawing showing where everything will eventually end up, but what about the processes in between? The materials will be delivered but how will they be offloaded? Where will the materials be stored whilst awaiting their final fixing position? What special facility will be needed for specialist equipment and materials? Will special handling and storage be required for such items as pressurised gas cylinders, flammable materials, pre-finished materials and so on (see Figure 10.8)?

The Operations Director decided a plan of action and with team meetings any amendments or adjustments could be made. On this particular project only 22 weeks had been allocated for all pre-production work. During that time the plan of approach had been formulated, but procurement is still required. (Procurement means to obtain, arrange and programme such things as labour, materials and plant.) A list of suppliers and specialists had been assembled over a long period on other projects. Where companies have previously performed well and supplied materials and plant they will be asked to tender for the new project. Those who have a history of, say, poor health and safety

Figure 10.8 Specialised storage facilities for bottled gas.

issues will not be invited to tender. Occasionally, however, price dictates that troublesome contractors will be used again. All trade contractors are continually monitored for health and safety with attractive incentive schemes to promote good practice (see 10.10, Health and Safety on and off site). The principal contractor then recommends specific trade contractors to carry out the work. Contracts for the supply only of materials and/or the supply and fix are then written out between the client and the trade contractor. Mace in the capacity of principal contractor were contracted as advisors and concerned only with the logistics and outcome of the project. They did not order or supply any of the materials, plant or labour other than that for communal use. In standard building contracts, by contrast, the main contractor would be responsible for all procurement.

Planning plant and site accommodation can be very costly if it is not done well. For example, how many site workers will there be on site in the first week? What jobs will need to be carried out? When Mace took over the contract the final stages of demolition were taking place. The first issues were:

- A large ground slab and virtually flat site that the new structure will cover.
- Where will the site management team be located? If they are located on site they will have to be moved whilst the substructure work is carried out.
- Where will the site welfare accommodation be placed? A similar situation again.
- Temporary services will be needed: water, electricity, sewage/drainage, telephones, broadband connections.
- What voltages and amperages will be required?

- What about food and drinks? Even at a very early stage on site the workforce still need plenty of tea and food. Site workers are often on site by 7.00am, meaning they will probably have been up since 5.00am to travel to work every day, so unless there is a very local café there will be a problem. In London, like many major towns and cities, the local café owners no longer want site workers. The very nature of site work means dirty, muddy or generally messy clothes most of the day. Unless the workforce changes their clothes every time they stop for tea you can see the problem. A site canteen is required. What about hygiene? The workforce may have to wear dirty clothes, but they still need facilities to wash their hands etc. and eat wholesome food. Therefore the site canteen has to ensure food is fresh, kept refrigerated or frozen as required and the food preparation area is constantly clean.
- What about access (termed 'ingress' and 'egress')? Where would access be placed on site? How will it be controlled? Should there be two alternative access points? What about emergency access if the main point is being used? It is not practical to only have one access on and off the site therefore the planners need to look at the initial needs and compare them to future needs as the site progresses. Logistics are similar to the game of chess. It is important to think several stages ahead and try to foresee any possible issues or difficulties.
- Selection of plant. Although the contract is for Construction Management and Trade Contracts, there is plant that will be shared over the duration of the contract. For example, tower cranes. Where should it or they be sited? How many are needed? How will they be erected – fixed or on rails? If fixed, what will they be fixed to? Eventually they will have to be removed, but how and when? Cranes are expensive items of plant. Unlike most plant they are susceptible to wind. At ground level it may appear to be a calm day but at 100 m or more on a relatively small sectional tower, wind can be a problem. Where will the energy come from to power the cranes? What about the environmental issues?

So far we have looked at the pre-production issues:

- how the budgets are formed enabling the operations director to know the maximum amount of money that can be allocated for each part of managing the project.
- site set-up considerations – the logistics
- the welfare for the workforce
- plant selection.

We now look at some of these in more depth:

10.8 Site set-up

The scenario: a small island site in a busy city centre and a matter of a hundred metres away the largest residential area in London, The Barbican. The other three sides are surrounded by multi-storey office blocks full of people trying to do business. The surrounding roads are relatively narrow and constantly busy feeding traffic through the city centre. Delivery vans and lorries constantly wanting access to the surrounding buildings. Parking bays and loading bays line both sides of the surrounding streets. The footpaths burst with pedestrians at the start, lunchtime and end of the working day, and still flow with business people all the day through.

The solution

- Visit the site and compare with the contract documents.
- Gather information about who will be on site, when and for how long.

From the information gathered, a graph was produced showing the duration of the site works on the 'x' axis against the numbers of workers on site on the 'y' axis. At this stage it would be impractical to wait until each trade contractor can supply the number of workers or site operatives or plant and facilities required. To overcome the problem the planners list the main activities to be carried out from the contract documents and then estimate the number of personnel that would be required based on previous contracts. It is not an exact science, only an approximation to plan for site accommodation etc.

As the structure is going to cover the entire site there is no space for storage, everything has to be programmed. The planners have to visualise each part of the project and how it will relate to later stages. It would be inconceivable to erect a crane, for example, and then find it is in the way later – or would it? Tower cranes are heavy, expensive pieces of plant that require a sturdy base to attach the tower. Typically, static tower cranes are erected where the lift shafts will eventually be. However, 'hook coverage' as it is known needs to be based on the needs of the site. On the Ropemaker site luffing cranes were selected. With a good lifting capacity of 12 tonnes at 23 m and 5.2 tonnes at 40 m radius the cranes could provide excellent hook coverage for the site.

Another issue with cranes is height; the taller the mast, the more susceptible the tower is to swaying in the wind. (Standing on the roof deck of the structure I was still lower than the crane cab. The weather was hot and at ground level the air was still, yet at about 100 m higher the cab swayed gently from side to side without any loading. To minimise the sway the masts are tied to the structure, enabling taller towers. You need to be very happy with heights to work as a tower crane driver.) All the tower cranes were fitted with anemometers that constantly monitor the wind speeds at the top of the cab. If the wind is too great or becomes too gusty the crane cannot be used. All the data is logged and can be checked if a trade contractor claims that wind prevented their part of the contract from being completed on time.

The crane towers are sectional and can be added to, thus jacking themselves up in situ; a demanding technique that requires the crane jib to raise the tower insert ready for the jacking process (see Figure 10.9). The planners need to programme when the jacking can take place as it will mean the crane will be out of action during the process and the testing period.

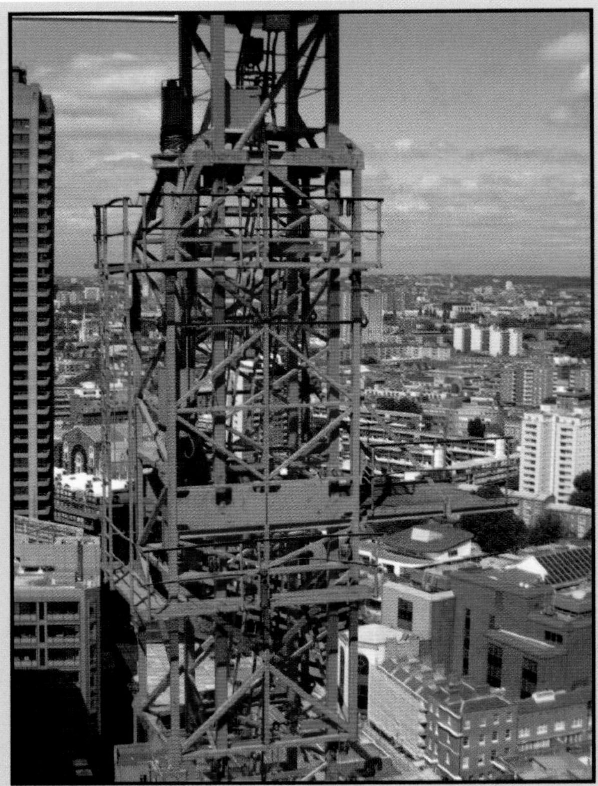

Figure 10.9 Tower crane mast jacking unit.

The planners at this stage had taken over a site that was at ground level and required major sub-terrain work. The old concrete slab floors and associated foundations had to be removed whilst the surrounding roads and buildings continued to be supported. The structural engineers had designed a new wall of concrete formed by secant piling held back by temporary steel works, but the location meant that special considerations were needed:

- *Protection:* Now the principal contractor has taken possession of the site, he becomes responsible for site security. The most effective method is to erect close board hoarding on the curtilage of the site to a minimum height of 2.4 m. Before erecting the hoarding a licence must be obtained from the Local Authority and a copy must be displayed for public view (see page 356 in *Building in the 21st Century*). As the site was basically a solid slab of concrete, holes would be needed for the hoarding and gate posts. The problem of noise would have to be addressed.

  Protection for the surrounding area is also important. To ensure the public footpaths and roads are not left in a permanently damaged state a deposit is placed with the Local Authority for every square metre surrounding the site. Set by the Council, the deposit runs into several thousand pounds which has to be set aside at the outset from the principal contractor's sum. The same applies to the removal of any street furniture such as lamp standards, parking meters, bollards etc.

Before any work on the new contract took place, a full visual inspection of the sewers around the site was carried out. It is important to know what condition the sewers are in as they may be partially blocked or cracked by a previous contractor. A visual inspection using specialist cameras enables a true record of the condition prior to commencement on the site. A new foul water connection had to be made for the site accommodation; kitchens, washrooms, showers and toilets, and surface water connection via an interceptor for drainage from the site. All surface water from the site passes through settlement tanks and an interceptor to ensure no silt or debris can pass into the mains sewer. Plant such as the vehicle washers used recycled water not connected to the drainage system.

- *Temporary site services: electricity:* To establish the electricity demand for the duration of the site the planners prepared a histogram based on predicted demand and when it would be required. For example, the tower cranes require 400 V at 300 A, whereas the hoists require 80 A. A Figure of 250kVA was calculated and a temporary electrical service brought on to site and housed in a brown GRP service room. All electrical supplies were taken off via armoured cables to high voltage plant or via transformers for distribution around the site.
- *Water:* Running in a temporary water supply is more straightforward than electricity. A temporary connection was made to the adjacent water main with a water meter and stopcock via a communications pipe. On site there was a manifold of connections in blue polythene and polypropylene (drinking water quality) plastic pipes to the kitchen, washrooms and other site needs. To comply with the Water Regulations anti-siphonage valves must be fitted on all site taps. Using plastic pipes has the advantage of flexibility and ease of making connections. All floors had temporary potable water supplies to serve drinking water fountains, emergency shower heads and large water cisterns to enable contractors to wash equipment such as trowels, buckets and floats. The use of plastic cisterns on each floor eliminated the need for contractors to wash tools in the washrooms.
- *Noise:* Breaking out old mass concrete is a long and very noisy operation. The work could not start early in the morning due to the Barbican housing only metres away. The adjacent office workers would not appreciate continual heavy breaking either, therefore the planners decided to diamond cut the concrete thus creating minimal noise.
- *Dust:* Cutting old concrete creates large volumes of dust therefore the diamond cutting machines have a continual water feed that keeps the cutting heads cool and eliminates dust.
- *Heavy plant for removal:* The broken/cut concrete required tracked 360° machines to lift and load haulage lorries. To enable continuous access to the work area and road haulage 'muck clearance' lorries an earth ramp was left at the corner of the site.

Figure 10.10 Early site set-up.

- *Mud:* Mud on the road is dangerous for other road users and if washed down will clog the main drains and road gullies. The planners decided a temporary road would be needed to enable the road haulage (termed 'muck clearance' or 'muck away') to remain virtually mud-free whilst the 360° machines loaded them. The roadway (see Figure 10.10 top right hand corner) incorporated a washing area where every lorry had high pressure water jets to clean the underside and wheels plus a road marshal ensured there was no debris trapped between the twin rear wheels. The used water returned to a settlement tank where the mud settled out from the water enabling the water to be re-used for cleaning. This is both environmentally friendly and prevents the main drainage/sewers from silting up. Mace had street cleaning vehicles on standby however in the event there was nothing to clean.
- *Site accommodation:* This could not be placed on site as the whole area would be used therefore the planners applied for a licence to temporarily close the only public footpath adjacent to the site and erect a gantry over it. Public footpaths are not designed to take high imposed loadings, they are designed to support 5kN/m^2 whereas the road will take 10kN/m^2. The solution was to erect the frame of heavy steel sections on a raft of mass concrete to provide support for the multi-storey village of site accommodation (see Figure 10.10). The public footpath was re-opened beneath the Portakabins with heavy wooden baulks, handrails and lighting to enable continued safe footpaths for pedestrians on the same side of the busy road (see Figure 10.11).

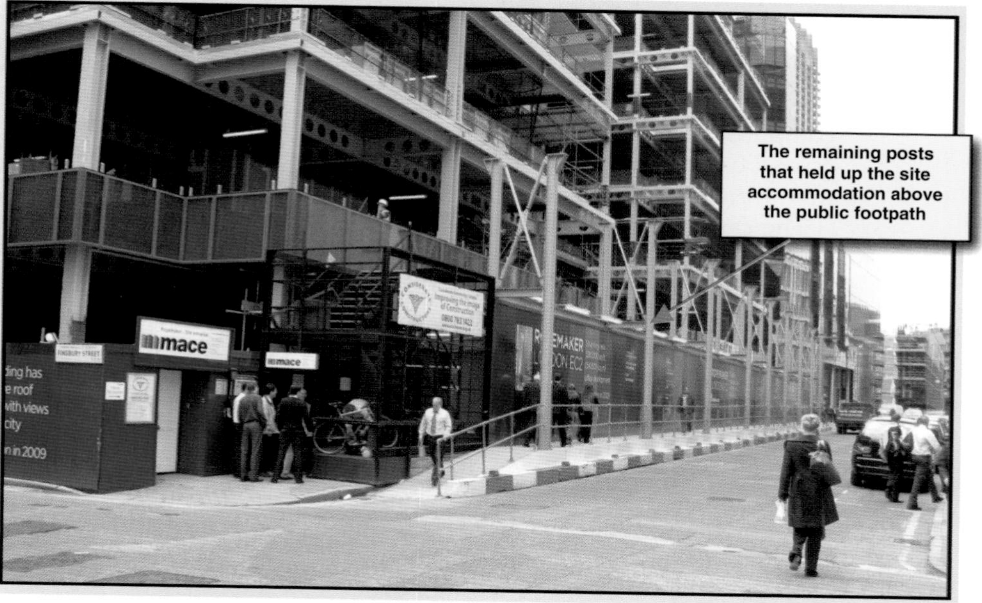
The remaining posts that held up the site accommodation above the public footpath
mace
mace

Figure 10.11 Pedestrian safety is a very high consideration.

Site accommodation is a legal requirement contained in the Construction (Design and Management) Regulations 2007 under Schedule 2. Although the requirements are defined as 'suitable' and 'sufficient', there are no prescriptive numbers laid down therefore it is a matter of judgment both on health and safety, and economic grounds. What is 'suitable' or 'sufficient' though? It could be considered as 'what is reasonable?' A useful guide as to tangible facilities against numbers of people can be found in the Metric Handbook.

- *Site boundaries:* An application had to be made to Islington Council for permission to suspend the parking and loading bays adjacent to the site. During the suspension the principal contractor has to pay a set sum per bay per day in compensation of lost revenue to the Council, all of which will be eventually passed on to the client as part of the contract sum. Therefore suspension periods are kept to a minimum.

Site accommodation

On-site accommodation comprised offices for the:

- operations director and costing manager
- site planners (on-site planners only, the others were based in local office accommodation)
- senior construction manager plus the members that make up the on site teams.

In addition:

- *The canteen:* Mace has a policy that where there is a canteen it is for all employees. The facilities are exceptional: large wide-angle flat screen TVs, computer terminals to enable site operatives to go online during their breaks, battery chargers so that the battery-powered tools can be charged up during the break times.

  The canteen is a good meeting place and Mace has an all-inclusive employment policy. On the Ropemaker site there are over 25 different languages spoken. All site workers have to attend a full induction programme prior to working on the site. You may be very skilled at the type of work you are doing on site, but it is not practical to have an interpreter with you all day. To act as a reminder of what the various Health and Safety signs mean there is an electronic touch screen in the canteen that will translate everything into several different languages (see Figure 10.12).

  Planning how to get the food into the kitchens and canteen had its own issues. To vertically transport literally tonnes of food and drink per week special hoisting and offloading facilities had to be installed on grounds of hygiene and availability. It would be impractical to carry the provisions up stairways or to stop the site work and use the materials or personnel hoists.

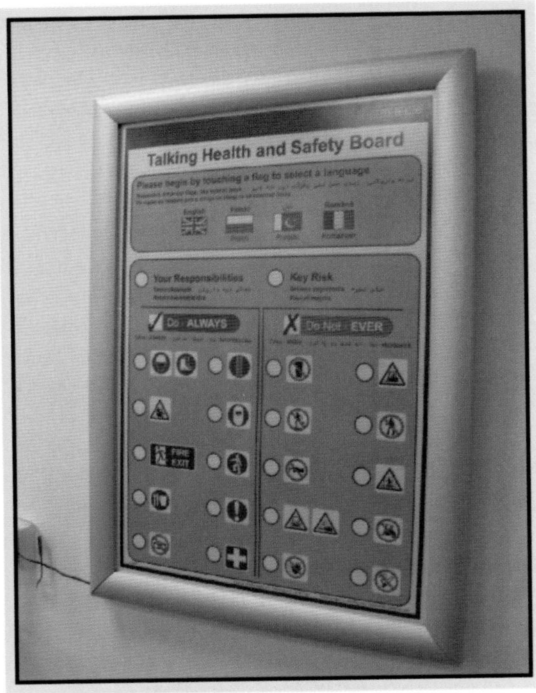

Figure 10.12 A multilingual health and safety board.

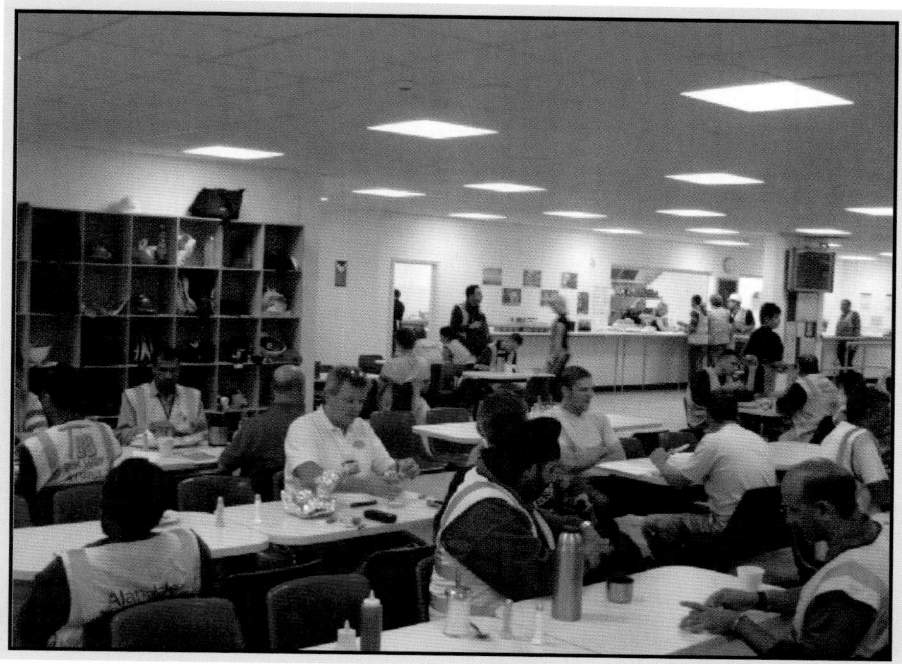

Figure 10.13 Excellent canteen facilities.

For those who want to take their own food to work there are 'serve yourself' kitchen areas with microwaves, sink units etc. plus seating, tables and storage areas for PPE, and portable tool battery charging whilst at break.

As the site progressed, all the initial accommodation had to be removed and new temporary accommodation installed. Canteen facilities were moved from the stacked Portakabin accommodation to the lower floors of the semi-finished part of the main block. The new canteen is large enough for the 600 on-site workforce constructed in studwork walling and brightly lit suspended ceilings and clean vinyl flooring (see Figure 10.13). For workers at the top of the building another smaller kitchen has been set-up ready for use. It can also be used as a back-up if maintenance is required in the main kitchen. The logistics of moving an entire kitchen with all the provisions, equipment including ovens, freezers, refrigerators and seating areas, plus all of the offices, wet rooms, first aid rooms etc., without closing the site had to be completed over a weekend.

- *Wet rooms/mess rooms* are needed as places of refuge in inclement weather periods. In contrast with the old wooden site hut, a few benches and a propane heater up one end, the Mace facilities are excellent. Purpose-built fitted-out rooms with benches to sit on whilst changing and then all the clothes and possessions are put on to a hanger/tray and given over to the security attendant (see Figure 10.14).

Figure 10.14 Secure mess changing rooms.

It serves two purposes:

1. A secure area for clothing and belongings required for the workforce to travel back and forth to work. The coat hanger frame ensures the clothes are not creased.
2. The changing areas are open plan and kept very clean and more akin to a sports complex changing room rather than a construction site.

Toilet areas and washrooms including shower cubicles and hand washing facilities are regularly cleaned throughout the day.

- *Security rooms* are required where important site documents can be kept, along with keys for secure compounds and a log for all those entering the site. Apart from security of the site the register is important in case of an emergency such as fire or evacuation. The log/register provides a permanent record of who has been on the site; the person arriving, the company or organisation they represent, who they are visiting and the time they arrived. An electronic tag is issued, strengthening security by allowing access through a turnstile system. When the visitor leaves the site, the tag is returned and the person signs out. The system is effective and ensures only bona fide people are on site.

- *Emergency facilities/first aid:* On a large and potentially dangerous site the facilities need to be much more than a green box with a white cross on it. Hopefully, with extensive site training, toolbox talks and lots of supervision accidents, will be minimalised. However, accidents by their very nature will happen (see 10.10, Health and Safety on and off site). The Ropemaker site has a very well equipped first aid room with trained first aiders throughout the site.

  Fire is always a major concern. According to the HSE there are on average 11 construction site fires every day. Construction sites often contain high concentrations of flammable materials awaiting fixing. For example, many years ago several large packs of expanded polystyrene cavity wall insulation caught fire due to an accident, causing several tens of thousands of pounds of damage and lost time cleaning black smuts off finished work. The heat generated from stored burning materials can deform the structure requiring partial re-builds. Fire on site can be particularly dangerous during construction as compartmentation and fire stops/checks may not be installed.

  On Ropemaker a full fire/evacuation drill takes place every six months to ensure the procedures work. Appointed fire marshals are responsible for evacuation of specific areas and on a daily basis monitor and check firefighting and alarm equipment. Each floor has a bright red stand showing the emergency escape routes above a complete set of fire extinguishers for use on paper, liquid and electrical fires. Fire/evacuation sirens are strategically placed to ensure all workers on every floor are aware of the evacuation.

  Anyone carrying out procedures that entail the use of gas flames for cutting, welding or brazing, or involve cutting discs and angle grinders must have a 'permit to work' and a 'hot works permit' before the operation is carried out. In addition, fire guards are required and personal fire extinguishers suitable for the task being carried out (see Figure 10.15). Mace have permits for all procedures that must be completed and logged before any work is done (see Figure 10.16). The method statements and permits are reviewed daily to ensure any conflicting procedures do not take place. However, we are all human. What happens if, say, a lifting lug on a steel bracket is slightly in the way? The lug has done its job and the component is in position. The lug, though, is preventing another component from being fixed. What do you do? Use an angle grinder and cut the lug off and continue fixing your component, or stop work and complete a 'hot work permit' and obtain a fire shield and personal fire extinguisher? If the former happens and is spotted it will become a recorded incident on the 'SHE' programme. If several incidents against a trade contractor are logged, further action is taken with the ultimate action of removal of the offender from the site.

  Alternatively, if the latter takes place then the trade contractor will have to stop work until the paperwork and equipment are in place,

Figure 10.15 Fire extinguisher points on all fire routes on every floor.

> adding time to the job being carried out. It might appear petty, but if something went wrong it would possibly be classed as 'negligence' and not an accident, as the procedures were not being followed. With SHE software the trade contractor management are automatically made aware of the incidents, enabling either further safety training or additional more experienced help on site. The idea of incident reports is not penal but to promote safe working.

Although at this stage the number of personnel on site was still minimal the planners had made provisions for several hundred trade contractors as the site progressed.

FM_032 - Hot Works Permit Form

Originator: John Hanley

Hot Works Permit				**Ref:**	
Project:	Permit No:	Issue Date:		Time:	
Location:	Company:	Expiry Date:		Time:	

Section 1	**Application:**	To be completed by supervisor directly responsible for the works to be undertaken
Type of Hot Work:	Key Fire Risks:	

Controls/Precautions

1. Area shall be cleared of combustible materials ☐
2. Non-combustible screens/sheets shall be erected ☐
3. Fire-watcher is required and shall be provided ☐
4. Local fire extinguisher shall be provided ☐
5. Flash-back arresters shall be fitted to LPG cylinders ☐
6. Containers shall be provided for used welding rods ☐
7. Contain welding dross, slag and sparks (Fire Retarding Material) ☐
8. Other Precautions ☐

Failure to comply with any of the ticked precautions will result in hot work permit being suspended.
I hereby declare that the above precautions shall be put into effect prior to the commencement of operations.

Signed: ________ Name: ________ Date: ________ Time: ________

Section 2	**Issue:**	To be completed by Mace Fire Officer (or Mace Manager)

Signed: ________ Name: ________ Date: ________ Time: ________

Section 3	**Clearance:**	To be completed by supervisor directly responsible for the works to be undertaken

1. Hot Works at the above specified location have ceased ☐
2. The area has been inspected for possible fire risk ☐
3. LPG Cylinders have been removed to a safe location ☐

Signed: ________ Name: ________ Date: ________ Time: ________

Section 4	**Cancellation:**	To be completed by Mace Fire Officer (or Mace Manager)

THIS PERMIT IS HEREBY CANCELLED

The Fire Officer has been notified of cancellation ☐

Signed: ________ Name: ________ Date: ________ Time: ________

Section 5	**Inspection:**	To be completed by Mace Fire Officer (or delegate)

Area inspected within **one hour** after cancellation ☐

Signed: ________ Name: ________ Date: ________ Time: ________

FM_032	Rev 02	Oct 00

m|a|c|e

Figure 10.16 Hot Works Permit form.

Mace strongly supports and in many cases surpasses the requirements of the 'Considerate Constructors Scheme', and the 'Construction Code of Practice' as set out by the London Borough of Islington's. The full document can be downloaded from their website. Many Local Authorities have their own versions of the code but like all codes they are not legal documents. However, many of the issues raised are based on compliance with legislation. The method of presentation makes for easy reading and a useful document: www.islington.gov.uk/DownloadableDocuments/Environment/Pdf/code_of_construction_booklet.pdf.

Construction can be a very noisy and dirty operation, but as indicated above, with suitable planning the problems can be kept to an absolute minimum. Knowing what is going on is perhaps one of the simplest issues that can be resolved by prior communication. Mace send out a monthly letter to all the neighbours of the site letting them know what their plans are and any local issues. For example, some very large or heavy plant may have to be brought in, erected or dismantled at the weekends or after the evening rush hour has ceased. By letting the neighbours know in advance, this enables them to make other arrangements; to go out for the day, for example, or, if they are like me, to go and watch. Every newsletter has names and telephone numbers to enable any specific concerns to be aired immediately, plus there are regular meetings with the site management, Local Authority and representatives from the Barbican.

Although the site was not within the City of London boundary, pollution does not stop there so Mace also liaised with their Environmental departments. Irrespective of how large or small a development is, keeping the neighbours informed prior to any noisy or disruptive work is always better than having to deal with complaints after the event.

Where possible the planners at the outset programmed all noisy work to be carried out at specific times during the day.

- Weekdays:
  - Site opens 7am with no noise before 7:30am.
  - Quiet periods 10am – 12 noon, and 2pm – 4pm.
  - Site closes 6pm.
- Saturdays:
  - Site opens 8am and no noise before 9am in consideration of the residents of the Barbican.
  - Site closes 1pm.

All trade contractors were informed of the two quiet periods during weekdays and limited noise on Saturdays. Although control over the site operations was maintained it could not extend to road works by utility companies. This created issues with some neighbours who thought the noise was coming from the site and accusations of breaking the noise curfews.

To enable work to continue throughout the day, quiet systems of working, mufflers on noisy plant or sound screens were used. First thing every morning a 'team meeting' with representatives of the trade contractors working on site and the construction management was held. The advantage of the daily meeting was that up-to-date information could be passed on to everyone concerned such as if delays were being experienced or materials could not be moved then alternative measures could be made. Issues such as storage space and plant downtime for maintenance can be overcome before they exist, a very important part of the planning programme.

Noisy operations and polluting machinery were kept to a minimum. The planners decided that mains electricity and bottled gas would be used where possible. Obviously some plant has to be powered by diesel but kept to the minimum.

Each contractor had to provide a method statement outlining the procedure and safety issues they will work to. That included the number of workers involved and who will be in charge of the operation. For example, the concrete floor slab will need to be broken through before the secant piling can commence. Noise and dust as previously mentioned would be a problem. If pneumatic plant is used economically there will be noise from 7 am when most sites start through to 6 pm at the end of the day. No one will agree to that in a built-up area and it would not meet the quiet time assured by the principal contractor. An alternative to pneumatic breaking would be to use chemical splitting where deep holes are drilled into the concrete and an expanding chemical mixture literally cracks the concrete. Apart from being slow, it is also very expensive. The system chosen was diamond cutting. Water-fed diamond-coated saws are relatively quiet. They create virtually no dust as the continuous water feed keeps the blade cool and soaks the cutting area as the cutting takes place.

To remove the waste concrete heavy lifting plant is required. There are several different types ranging from HIAB self-loading tipper lorries to front-loading shovels, either tracked such as a drot, or articulated pneumatic tyred machines. To pull up the cut concrete another piece of plant would be required, therefore a tracked 360° back actor was used (see Chapter 9 in *Building in the 21st Century*). The machines have the power to pull out the cut concrete, slew round (rotate) and load waiting lorries. To ensure noise is kept to a minimum, the site operations were planned around the two quiet periods per day. As you can see, noise issues can affect the programming of a site significantly. The site operatives still need paying during the quiet periods which all has to be taken into consideration when tendering for the work.

10.9 The ground works

Before the basement levels could be excavated ground support had to be installed. A 17 m deep secant piling system comprising 270 piles had been chosen by the engineers. Firstly a series of concrete piles are augured into the ground then steel reinforcement cages are lowered in before placing the concrete. (An alternative method is to place the concrete in the pile shaft and lower the steel reinforcement cage in before the concrete begins to set. However, accuracy of positioning is essential to allow the auguring of the male piles without hitting the reinforcement of the female piles). The first piles are classed as female piles cast in slow-gain concrete spaced about 200 mm less than the pile diameter apart. The initial curing takes place in 2–3 days and then the male piles are put into place, cutting into the female piles between 75–100 mm on each side and forming a homogeneous wall of steel strengthened concrete. A continuous beam is cast across the top of every pile. This is known as a pile capping beam and as it continues on all four sides it is also a ring beam. The site is sloping therefore a step was created to accommodate the change in height.

The next stage was to excavate over the site plan leaving one corner for removal of the spoil (excavated materials) and site access. To prevent the secant piles deforming and the soil support failing, long steel tubes known as temporary support were installed using mobile cranes. Spanning the width of the site and bracing each corner, the steel tubes enabled the full basement depth to be achieved (see Figure 10.10). When the formation level (the point where the existing ground finishes and the underside of the new works meet) was reached, the raft foundation could be placed. With literally tonnes of fresh concrete and steel placement, an armada of concrete transit lorries and concrete pumps placed 5,365 m^3 of concrete with 785 tonnes of steel reinforcement. To pour such a massive amount of fresh concrete takes careful planning. Cement is exothermic, meaning it produces heat from a chemical reaction. The raft comprised a series of slabs joined together with continuity bars and water stops. The first slab became the base for the first tower crane on site. In the lower basement each reinforced slab 1.5 m thick × 15.0 m × 12.0 m required 270 m^3 of concrete. The ready mixed concrete was brought to the site on a fleet of transit vehicles, each carrying 8 m^3 equalling 34 lorries. Say each lorry had an off-load time of 15 minutes, how many hours would it take to just deliver the concrete? Answer:

$$= \frac{\text{Total number of cubic metres in each slab} \times \text{the off-load time per lorry in hours}}{\text{number of cubic metres of concrete per lorry}}$$

$$= \frac{270\,\text{m}^3 \times 0.25}{8}$$

= approximately 8.5hrs.

That is assuming the concrete is poured continually as a rate of approximately 2 minutes per cubic metre. Planning the labour requirements to enable operatives to have their breaks and the timing of the transit lorries plus adequate wash down time at the end of the day was crucial. All concrete pumping had to be complete by 4:30pm to enable the plant to be washed out ready for the next day, therefore a prompt start at 8:00am was required.

To deliver and place the raft foundation alone required 20 days at the rate shown above. However, by doubling up on pumps and labour the time was reduced by 3 days.

The site at this stage has relatively few site operatives (trade contractors) on it, though planners and management team need to be on site overseeing the work.

Mace chose two groupings:

1. temporary site accommodation for the site works
2. rented accommodation/office space for the planners and costing sections in a building close by.

10.10 Health and Safety on and off site

There are several pieces of legislation that cover the construction industry. Not all apply to every site, but you can work on the basis that most will. They are:

Health, Safety and Welfare at Work Act 1974

Under the Act the employer is responsible for implementing Health and Safety throughout the whole project. British Land Plc. are the employers therefore ultimately responsible for compliance. Every employee is responsible for their own health and safety and those around them. According to the Health and Safety Executive the construction industry employs at least 7% of the working population in Britain but accounts for 16% of the major accidents, 25% of which are fatalities. So health and safety is the number one objective. The principal contractor in collaboration with the CDM co-ordinator act as agents to the client, promoting good working practices and where necessary dismissing trade contractors who habitually break the safety rules.

Barry Beck is the health and safety manager on site, and also the senior construction manager. With several decades of experience working on major developments around the world, he daily tours the site monitoring all parts of the work. Each trade contractor has a person responsible for health and safety issues who meet on a weekly basis to discuss any non-urgent points. Urgent issues are dealt with immediately and recorded on the 'SHE' software programme.

As part of the research for this chapter I spent a day with Barry as he toured the site. Small issues such as paper packaging that had not been placed in the appropriate skip did not strike me as unusual but as he pointed out, a service engineer using a cutting disk above was creating hot sparks that although very unlikely could ignite the paper. Where a trade contractor needs to use a cutting wheel a 'hot work permit' is required. The idea is to ensure a risk assessment has been carried out for the operation and the appropriate safety procedures and PPE are being used. For example, one trade contractor has produced a small screen to prevent any hot sparks from a cutting disc or when working with bottled gas for welding, brazing or soldering escaping from the work area (see Figure 10.17).

The Manual Handling Operations Regulations 1992

Where reasonably practical the employer should avoid employees manually handling materials. Mace installed a steel gantry and a steel jib to offload materials from delivery vehicles. Hydraulic pallet trucks were available for movement over flat area and floor decks. Fresh concrete was offloaded by chute into concrete pumps negating any manual handling only placing and finishing. Several materials hoists were strategically placed around the access areas plus people hoists for site personnel to travel to various levels of the

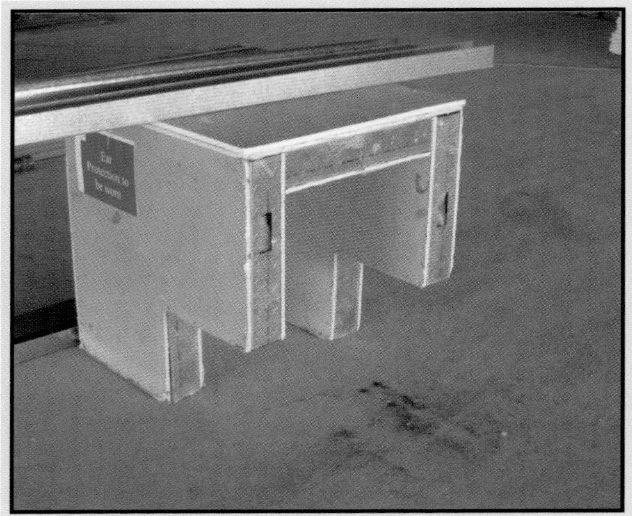

Figure 10.17 Purpose made hot works screen.

construction. Several tower cranes plus small crawler cranes and glazing robots further aided site lifting (see Figure 10.18). Services fixed to the soffits of each floor include heavy steel pipes and other heavy items such as fan motors etc. Specialist mobile hydraulic lifting equipment enables the items to be held in place whilst fixing. This is in contrast with the old traditional method of manual handling and propping up with make-do equipment. (The manual method is still used extensively on smaller sites and sites condoning bad practice.)

Health and Safety At Work Regulations 1992

This requires the employer to carry out risk assessments on any work that is to be carried out on site and ensure that control measures are put in place.

Management of Health and Safety at Work Regulations 1999 (Management Regulations)

The employer is obliged to carry out the risk assessments of the workplace and review them when necessary. Records must be kept where more than five employees are on site. Mace use an electronic system. The software, known as 'SHE', allows data to be logged regarding any health and safety issues including near misses. Any member of the construction management team can access the full data. Where specific 'hot spots' occur, for example one particular trade contractor having several health and safety issues recorded against them, the management of the trade contractor will be notified, enabling them

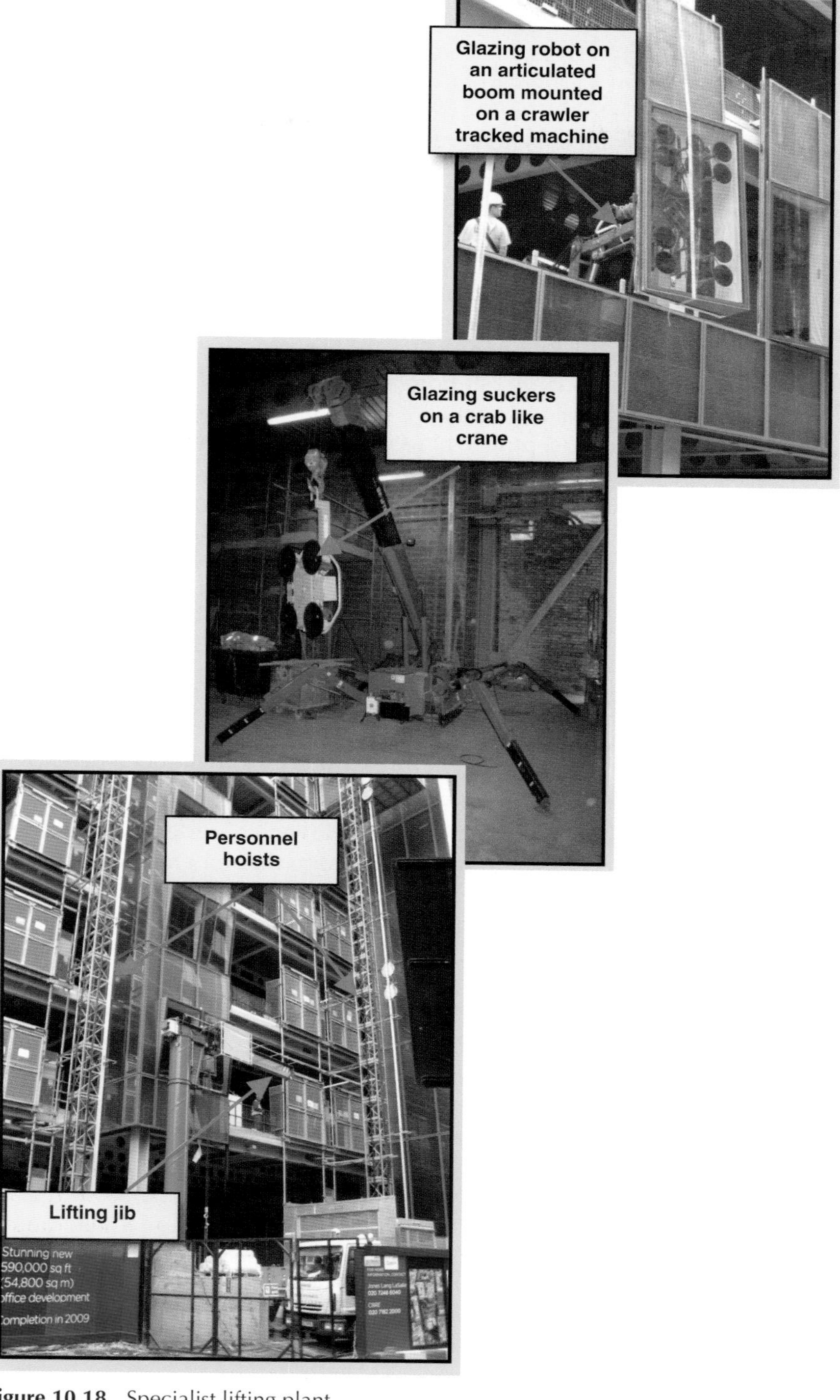

Figure 10.18 Specialist lifting plant.

to also monitor and remedy the situation. An example could be the trade contractor has presented an acceptable method statement to the principal contractor. The site operatives have been issued with PPE including eye protection in the form of polycarbonate face visors. The site workers are trying to install cabled services to the soffits in narrow corridors. The weather is hot and the location means there is no ventilation so the face visors continually mist over. The result is the operatives either work with the visors up or are continually having to stop and wipe them. An issue like that requires resolving therefore recording it ensures and confirms the trade contractor management are aware.

'SHE' software also enables a monthly 'Site Based Contractor Health and Safety League' to be produced. A list of all the on-site trade contractors is produced as a league table against 10 headings. Points are awarded for positive action where site management have recorded good safety practice beyond the expected levels. In contrast, points are deducted for such issues as not using appropriate PPE or not carrying out safety procedures. Other headings include good housekeeping where contractors keep their areas clear of waste and work in a tidy manner. All 10 categories are totalled up and designated as 'Excellent' down to 'Appalling/Very Poor'. The league is a good incentive for health and safety and promotes good practice. In return, monthly prizes are awarded to contractors and individuals in the form of vouchers to spend in high street stores and restaurants. The scheme is financed by the client and is very successful.

Work Place (Health, Safety and Welfare) Regulations 1992

Specific issues such as temperature, lighting, ventilation, cleanliness, room dimensions etc. only apply to the office workers on site. It would not be practical to have such regulations for the site. However, under the CDM Regulations the employer must where practical ensure that all employees have adequate lighting, ventilation and reasonable temperatures to work in. That may be achieved by space heaters on floors where the external envelop has been completed. (Particularly important where decorative finishes such as paints, wall coverings and floor coverings are being finished off.) Also under the Work Place Regulation there is a requirement for a rest area for smokers. Although smoking on site is not allowed for safety reasons, Mace have put aside an area where site workers who need to smoke can sit. It overcomes the problem where smokers would have to check off site and stand on the street to smoke. Whether smokers should be allowed time to smoke at work is something for debate.

Provision and Use of Work Equipment Regulations 1998 (PUWER)

In December 2002 the regulations included the suitability, maintenance and inspection of mobile work equipment. During one of my visits to the site a new crawler mobile crane had just been delivered and all operatives who would be using the new machine received onsite training (see Figure 10.19). As many

Figure 10.19 Specialist training for all new equipment.

of the site workers have come to London from other countries (25 different languages are spoken on site), interpreters ensure everyone has a full understanding of the training.

Lifting Operations and Lifting Equipment Regulations 1998 (LOLER)

This requires any equipment used for raising or lowering people or materials should be specifically designed for the purpose and regularly maintained. That includes boom lifts (commonly known as 'cherry pickers') and scissor lifts. The acronym MEWPs (Managing mobile Elevating Work Platforms) is particularly important as they are susceptible to overturning when used on uneven terrain. On Ropemaker site the MEWPs have wide wheel bases and operate on flat concrete floors. Where required 'out-riggers' (metal adjustable tubes acting as raking props that increase the base area of the machine, giving more stability) are fitted. They are commonly used by the services engineers who spend most of their time installing pipes, cable looms and ducting to the soffits of each floor.

Work at Height Regulations 2005

These also relate to LOLER, in that before any working at height task takes place a full risk assessment is carried out.

Mace chose a purpose-made edge protection system fitted to the steelwork prior to the floor decks being cast. Consisting of posts and two horizontal steel tubes welded at the joints, the edge barriers cannot be dismantled or adjusted, thus ensuring a constant barrier. When the external walling is fixed the barriers are disconnected then cut up and sent as scrap metal for recycling. This is both environmentally a good use of material and a very effective cost-efficient method of on-site edge barriers.

Site operatives such as the steel workers and crane maintenance engineers all wear full harnesses with retractable lanyards attached to their backs. The lanyard attachment should be located near the shoulder area as opposed to the small of the back. If the lanyard is not fitted correctly and the operative falls, the sudden jolt could cause back injury. Falls and trips account for the highest proportion of site accidents.

In addition to edge barriers, all the floors perimeters have large drag nets draped from the soffits to prevent any gusts of wind blowing light materials off the building.

Delivery vehicles may not be thought of as working at heights, but standing on the back of a flat-backed lorry is still a very dangerous height to fall from. When flat back articulated lorries arrive on site they are fitted with 'clamp on' edge barriers before any unloading (other than by fork lifts) takes place. Edge barriers are not always practical though. For example, steel beams, stanchions and other long and heavy members will be lifted by the cranes. The long length of cable from jib to the hook and chains, plus possible wind gusts mean the lift may move about. Each lorry parks adjacent to a loading platform designed at the same height as the flat back loading bed of the lorry.*

Alternative methods are mobile catenaries or static jibs (see Figures 10.20 and 10.21). Fully trained operatives known as a slinger signalmen are responsible for attaching chains, lugs or webs to the materials to be offloaded and guide the crane driver from the ground. If in the unlikely event he or she is either pushed off the lorry by the load as it is lifted, or slips off the steel work, they will dangle from the line and not fall. If edge barriers are used instead of the harness the banksman could be squashed/crushed between the load and the edge barrier or pushed over the top. Some hauliers have nets that splay either side of the lorry bed. In the event of the banksman tripping or being knocked off the load they will fall onto the net. The Working at Height Regulation 2005 covers any work above or below ground level where a fall could cause personal injury therefore even step ladders are included.

Health & Safety (Display Screen Equipment) Regulations 1992 (Display Screen Regulations)

It is perhaps unusual for a construction site, however due to the number of personnel managing the site the use of computers becomes a major part of their working day. Large screen monitors and fully adjustable work stations, regular eye tests and the provision of spectacles where necessary, plus breaks from using the screen ensure all computer users comply with the regulations.

* A requirement of the CDM Regulations 2007: Vehicles, 37.(5) is that no person may remain on a vehicle whilst a loose load is being craned off unless there is a safe place for them. For example, grab and crane lorries often have a specific seat behind a bulkhead bar to operate the equipment whereas other vehicles have remote controls to operate from ground level.

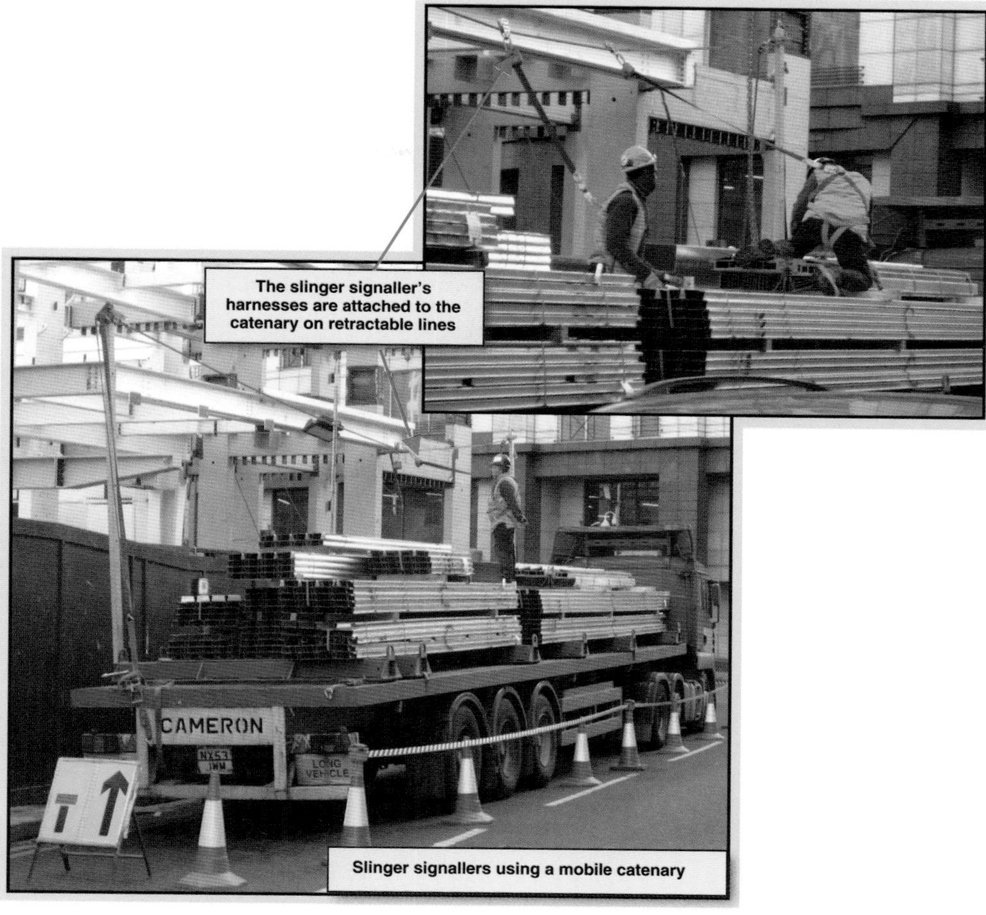

The slinger signaller's harnesses are attached to the catenary on retractable lines
CAMERON
LONG VEHICLE
Slinger signallers using a mobile catenary

Figure 10.20 Mobile catenaries for slinger signalmen safety.

Construction (Design and Management) Regulations 2007 (CDM Regs)

This comprises five parts:

1 Definitions and application of regulations.
2 General duties on construction projects.
3 Additional duties where the project is notifiable (lasting more than 30 days or involves more than 500 person days of construction work). Ropemaker project is notifiable. A form known as an F10 must be completed and sent off to the HSE or as from June 1st 2008 it can be completed electronically online as an 'EForm'. To read what information a site manager and CDM co-ordinator has to provide go to the HSE website and type in F10 in the 'search' box.
4 Practical requirement for construction sites.
5 Transitional arrangements and revocations.

The slinger signaller's harness is attached to the jib on retractable line
BYRNE BROS
SCANIA

Figure 10.21 Static jibs for slinger signalman safety.

The full Regulation can be accessed on the HSE website. There is also a very useful Approved Code of Practice (ACoP) that provides practical advice on how to meet the CDM Regulations 2007 available from HSE Books.

The Construction (Health, Safety and Welfare) Regulations 1996 (CHSW) have been transposed and absorbed into the CDM 94 and further updated with the CDM 2007 Regulations. For example there is no prescribed number of facilities for site workers. The CDM 2007 approach states that the requirements should be adequate.

Health and Safety (First Aid) Regulations 1981

This requires the employer to have adequate and appropriate provision for first aid on site. To ensure adequacy of provision the planners would check the risk assessments from previous trade contractors who have carried out similar work, method statements and company records from previous contracts. Where sites have used the 'SHE' software statistics of accidents, near misses, types of injuries will provide a good insight into the requirements of first aid on this site.

Reporting of Injuries, Diseases and Dangerous Occurrences Regulations 1995 (RIDDOR)

Where a employee dies as a result of an accident on site the HSE (Health and Safety Executive) must be informed together with the Local Authority and or Police. It also applies to any employee who is off work for more than 3 days due to personal injury or sickness. To record the information Mace use 'SHE' software which is a combination of a sophisticated database and spreadsheet system. All incidents including near misses, and anything related to health and safety issues are entered onto the programme. Full information about the software programme can be found on their website http://www.shesoftware.com/enterprise.html

Control Of Substances Hazardous to Health Regulations 2002 (COSHH)

Industrial exposure to different hazardous chemicals whilst at work should be kept to a minimum. In addition to the employer's a duty of care under the HSAWA 1974, and the CDM Regulations 2007, additional obligations under COSHH regulations exist. The principal contractor required a risk assessment and method statements from every the trade contractor *prior* to any materials being delivered to site. The information had to include any chemicals considered as 'hazardous to health' as listed in Part 1 of the Approved Supply List of Chemicals (Hazard Information and Packaging for Supply) Regulations 1994 (CHIP 2 Regulations). Materials such as adhesives, waterproofing coatings and paints may give off heavy fumes that require ventilation whilst being applied. Storage of the materials may present a fire risk or be explosive due to low flash points.

Industrial dermatitis can be caused by operatives not wearing the correct PPE or other operatives being unaware of the chemicals when cutting or shaping materials such as chemically treated timber products. Mace keep all of the COSHH sheets in file systems on site for ease of access.

Working Time Directive and Working Time Regulations 1998

European law limits the average working week to 48 hours. (Some companies due to the nature of their work can opt out, but it must be by written agreement of the employee. He/she can end the agreement without the consent of the employer. The law may state that, but the employer may terminate the employment although it then could be considered as unfair dismissal.) In addition, every worker should have 11 hours' rest in every 24 hours plus a minimum of 1 day off per week. Although the EU law applies to workers in the UK there are occasions when key workers will exceed the limits. For example if a specialist piece of plant is being erected or dismantled such as a tower crane it may take place after the site has closed. It would not be practical to stop part of the way through the process. The principle of the EU law is to prevent employers reverting back to the practices of the Industrial Revolution. It is interesting that the law does not extend to managers or executives, who do not have set working hours. They are considered to be able to adjust their working arrangements to suit the conditions. From April 1 2009 every employee is entitled to 28 days' annual leave (including Bank and other Public Holidays). Under European law an employee is not permitted to work and take payment in lieu of the holidays except on termination of employment.

Personal Protective Equipment Work Regulations 1992

Hard hats, gloves, protective footware, eye protection, ear protectors, etc. required when working on site should be provided free of charge to all employees. The principal contractor Mace ensures that all employees of all trade contractors attend and sign that they have completed a comprehensive site induction. In addition, where specialist trade contractors are working on site they receive extra toolbox talks prior to working.

REACH (Registration, Evaluation, Authorisation & restriction of CHemicals) 2007

The main objective is to control the use of chemicals throughout Europe. Chemicals used on site include paints, oils, solvents and cleaning materials, and cement. This new regulation is in addition to the COSHH Regulation 2002, so some overlaps do occur. Cement, for example, as a dry powder may not appear to be hazardous other than as dust but when in contact with water or moisture such as sweat, eye moisture, mouth and nose etc. the alkaline chemical content may cause burns.

An additional regulation 'The Classification, Labelling and Packaging of Substances and Mixtures Regulation' (CLaP) is expected in early 2009. Its objective is to further control the identification of hazardous chemicals and materials although it will not become law until December 2010 at the earliest.

Some oil manufactures produce bio-degradable oils including hydraulic oils for use on construction site heavy excavators and dumpsters. The advantages include the fact that any spillages or burst hydraulic pipes will not be hazardous or contaminate the ground. In contrast, conventional mineral oils are hazardous and any contaminated materials must be removed and disposed of as 'hazardous waste', i.e. not just dumped in the nearest landfill skip or covered over. In a similar way many paints and preservatives are now water-based and do not contain poisonous metal compounds such as copper sulphates.

All chemicals used on site must have a risk assessment carried out as to their use and application. Manufacturers will provide a Full Materials Safety Data Sheet (MSDS), also known as a SDS (Safety Data Sheet), many of which are available as downloads from their websites. If there are any doubts about the use of various chemicals, especially when used together, it is important to contact the technical departments of the manufacturers for advice *before* they are used. Some chemicals used in isolation present no hazards and are safe to use, but solvents may react with others to give off noxious or heavy fumes. For example, floor adhesives and waterproofing materials used in confined non-ventilated spaces such as stock rooms.

Electricity at Work Regulations 1989

On site there are various voltages used, from the heavy plant requiring three phase 400 V to the ovens in the kitchen on single phase 230 V. Transformers are used to reduce the voltage down to 110 V for mobile plant and as low as 5 V where the cables may lay in water. According to the HSE around 1000 recorded incidents a year involve electrocution at work, including about 30 fatalities.

The HSAWA 1974 Section 3 (1) requires every employer to ensure where reasonably practical that persons are not exposed to risk to their health and safety therefore wires or cables that have insulation tape around make-do joints must not be allowed. During the daily tours of the site particular notice of trailing cabling is made. Fire and evacuation are prime objectives therefore any cabling for lighting etc. of stairways is tied back neatly away from the stairs. Apart from keeping things neat on site it prevents anyone becoming entangled during evacuation (see Figure 10.22).

In contrast, not all sites have the same high standards though. Many sites have several cables carrying lighting to the stairways and extension leads to other floors cascading over the stair treads which are hazardous to normal stair use and could be lethal in the event of a smoky fire on site. Perhaps the worst site I have seen had ordinary screw-in light bulbs with no protection cages as a chain of lights more akin to Blackpool illuminations trailing down the entire stairwell.

Noise at Work Regulations 1989 (NWR)

Although there is a duty of care under the HASAWA 1974 the employer also has duties under NWR to ensure employees are not subjected to unnecessary levels of noise. Some trades use portable circular saws to cut sheet materials such as fire boards etc. The operatives doing the work should be wearing appropriate PPE which includes ear protection. However, what about other operatives close by such as service engineers installing electrical cabling? This was one of the issues that came to light during the tour with the health and safety manager/senior construction manager. Although programmed to be working on different parts of the same floor, the noise from the electric saw

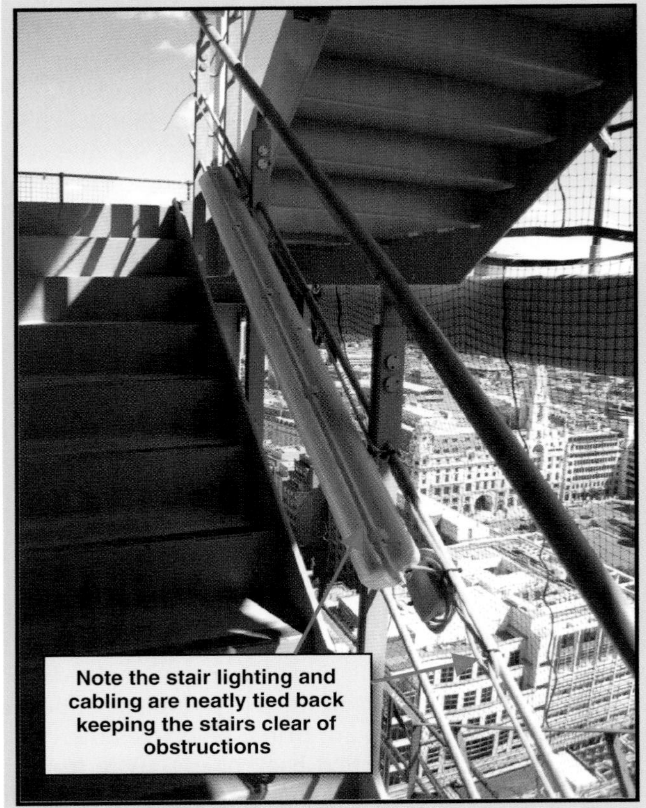

Note the stair lighting and cabling are neatly tied back keeping the stairs clear of obstructions

Figure 10.22 Neat and tidy emergency lighting to stair cases.

echoed loudly in the relatively confined area the electricians were working in. A decision had to be made: should one contractor should stop work whilst the other finished their work? Not really acceptable. Should the electricians be supplied with ear protection? It would solve the noise problem but may not be comfortable for the electricians especially as they need to speak to each other. A mobile sound screen placed around the operative using the saw could be the solution? Regulations must be looked upon practically and not as cold and fast rules.

Much of pre-site planning is based on experience from previous sites. The project may be different however the principles are more or less the same.

So far we have looked in depth at:

- the issues surrounding the site set-up
- the methods used to calculate what site accommodation will be required
- the problems of site layout with regard to access, including licences for temporary suspension of parking bays, footpath issues and retention of money set against potential highway damage
- noise and dust issues and how to limit the nuisance factor

- muck clearance and how to keep the surrounding highways clean
- site accommodation, where to place it and what facilities were provided
- planning for vertical movement throughout the site including the issues of supplies to a very large canteen
- wet rooms and washroom and shower facilities
- security on site and why it is very important
- first aid on site
- complying with Acts and Regulations associated with construction sites with comment of how the site work complied with the legislation
- the importance of liaising with the neighbours and keeping a close rapport with the Local Authority Environmental departments
- incentive schemes to promote good health and safety practice on site.
- the provision by the principal contractor of lifting equipment for both materials and personnel
- the method adopted used to excavate the basement area and lay the raft foundation
- the logistics of pouring the concrete for the raft foundation.

The role of the principal contractor could be compared to the conductor of the orchestra. Although not actually creating the work, he/she is there to ensure it all is in place at the right time and finishes as planned. The planners on Ropemaker used a computer programme which basically is a powerful spreadsheet that can be continually adjusted to meet any daily or even hourly changes to the construction programme. Asta Powerproject is one of several software solutions currently available. They work on a spreadsheet platform, but there are differences as can be seen on their website: www.astadev.com

10.11 Planning

The basics of planning are to create lists of the jobs, termed 'activities', that need to be carried out and then place them in construction order. Using a spreadsheet, prepare a series of columns headed as: Ident, task/activity, start date, duration (see Figure 10.23(A)). Then to the right of the duration column form groups of columns to represent days, weeks or months as appropriate for the contract. Weekends should be included. Then enter the data as required. For example, we know that as soon as the principal contractor has taken possession of the site then security should be undertaken on grounds of health and safety, therefore:

1. *Security:* Next would be to prepare the site for the new work. That might involve demolition of existing buildings as in the case of the Ropemaker site, or grubbing out (removing trees and large plants/bushes) or just the removal of topsoil from the proposed work area. For simplification we try to use as few words as practical; 'demolition', 'site strip' etc.
2. *Demolition:* Depending upon the type of project, site accommodation may be needed. This can be termed 'site set-up'.

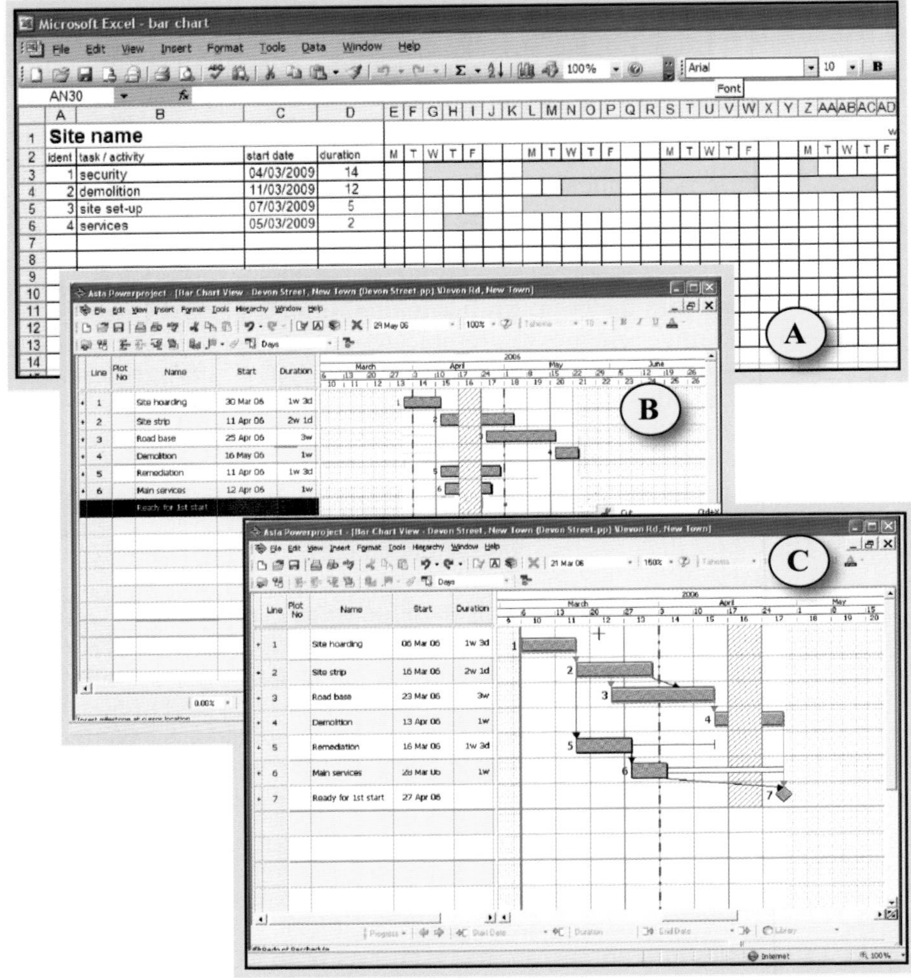

Figure 10.23 Electronic programming.

3 *Site set-up:* As you can see the list will be based on a logical progression of how the site will run. To help identify each task a number is assigned to each line. It is particularly useful when discussing issues over the phone. In the column next to the task a start date would be entered or the planned duration time. As the site requires hoarding on all four sides it is likely to take several days. If say it should take 14 days to erect the hoarding then '14' is entered in the duration column.

On the Ropemaker site Mace used the Asta Powerproject software (see Figure 10.23(B & C)). It is very easy to use and a very powerful tool. For example the data shown above would be entered in the appropriate activity column, but after entering the projected duration a date with a small pop-up calendar will be shown. It is an easy operation to change the start date and the software converts the 14 day duration onto a horizontal calendar to the right, putting in breaks for weekends. If there is, say, a bank holiday it is easy to cut the duration band and drag it to the next working day.

The activity list can be imported from a previous project and modified to meet the requirements of the new project. The durations are entered and then it is possible to drag the duration bands to the planned start dates. By attaching linking arrows from the duration bands a critical path programme can be shown. The green duration bands become outlined in red and numbered accordingly. Where activities are not subject to critical paths they remain green.

10.12 Critical paths

CPA (Critical Path Analysis) is a modified plan or programme showing the order in which the activities must be carried out. A simple example would be the foundations. The foundations normally go in first as without them the walls cannot be built. The ground floor would normally go in before the walls continue to the first floor level or plate level as in the case of a bungalow. The first floor would be built in before the walling continues and then the next floor or the roof and so on. However, so long as the main elements are constructed in the correct order the programme should proceed correctly. What, though, if the bricks to complete the walls do not arrive on time? Can the carpenters still fit the roof on the programmed date? The programme plan normally just shows what will be done, when it will be done and how long it will take. The CPA will show which activities are crucial and must be completed before the next activity can be completed. The manager can see which trade contractors will be affected and how the programme can be modified to overcome the delivery problem. The traditional method was to hand draw a bar chart or, as it was sometimes referred to, a Gantt chart. (Just to clarify matters, the Gantt chart differs from the bar chart in that the 'x' axis was calibrated by time. Henry Gantt wanted to show items against a timescale as opposed to the then convention of materials against quantities. Within 10 years of his modification the Gantt chart became the norm.)

The system of CPA otherwise known as networking or precedence diagrams was developed for major defence projects in the 1950s but has widespread use throughout industry and commerce. As with bar charts and all other planning tools, a list of tasks/activities must be placed in order of construction. A start date is needed as well as the duration period. When the data has been gathered it must be arranged in order of dependency whilst still in order of construction. There are several versions of writing the plan (see Figure 10.24), mostly involving circles, squares, rectangles or triangles divided into sections. The shape is not that important as long as it is constant throughout the chart. The bubble or more correctly 'node' contains information about the 'earliest start date' and 'latest start date' of the activity. Some techniques indicate the numerical order the activity will take which should coincide with the numbers used for activities in the bar chart.

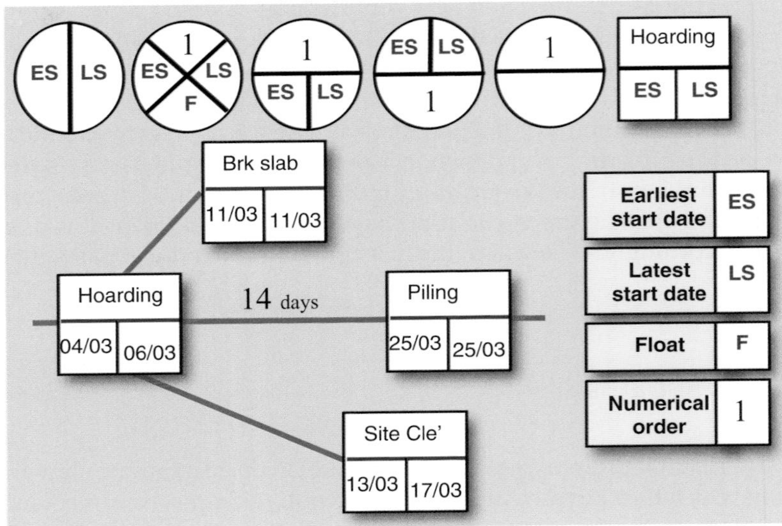

Figure 10.24 CPA programming.

When each activity has been identified with a node they would be set out on a chart, normally flowing from left to right, with the duration indicated adjacent to the line.

Example

We have a short list of activities that need to be arranged as a critical path analysis.

Activity order	Activity	Duration (days)	Earliest start	Latest start
1	Site security (hoarding)	14	04/03/09	06/03/09
2	Break out slab	9	11/03/09	11/03/09
3	Site clearance	5	13/03/09	17/03/09
4	Piling (275 piles)	20	25/03/09	
5	Form pile reinforcement cages	9	09/03/09	13/03/09
6	Break out pile caps	14	23/03/09	
7	Make pile cap formwork	5		
8	Form pile cap steel reinforcement	14		
9	Pour pile cap concrete	4		
10	Excavate upper basement	5		

Activity 1: hoarding – the site should be made secure before any work is carried out. However, if part of the hoarding had been erected would it be practical for the area adjacent to start breaking the concrete slab? If so, the heavy plant could be made secure at night on site and lighter plant chained and padlocked to the heavy plant for security. Any excavations must be protected with barriers under the 'duty of care'. The contractor must take reasonable precautions to prevent accidents even if the person is a trespasser.

The hoarding could be erected and the breaking out of the slab following on as work proceeds, therefore although 14 days duration and 9 days respectively, both operations could be carried out together. The hoarding start date of 4 March could not be any earlier as that is the date of possession, it could be up to the 6 March, though, as the site must be totally secure by 25 March when the piling rigs arrive on site.

The site clearance is programmed as 5 days. It cannot start until there is a reasonable amount of broken out concrete to warrant clearance. The 'muck away' contractor will need continuity of loads therefore the earliest start date would be 13th March and the latest based on 5 days for clearance would be 17 March. Breaking out the pile caps exposes the top of the reinforcement cage and enables clean concrete for the pile cap (a beam in this case) to be cast on. The exposed ends of the vertical rods of the pile reinforcement will allow the steel fixers to tie the beam reinforcement to them giving continuity. The shuttering carpenters will have to form a box that will hold the fresh concrete and steel reinforcement in the correct place therefore they need to build the formwork before the steel workers tie in the reinforcement cages and spacers. The timing can be staggered as the piling progresses therefore more information termed as 'float' can be added to the CPA. It then becomes more complicated. To help understand the planning process:

1 Take a sheet of paper and cut it into 50 mm squares.
2 Write an activity on the top of each piece of paper and a 'T' beneath as shown in Figure 10.24.
3 On the left of the 'T' enter the earliest the activity can start and on the right the latest.
4 Now arrange each activity node so that the main flow of activities goes left to right and place on a large sheet of paper.
5 Draw lines linking each node in the direction the activity must take place. There may be several lines of progress running parallel culminating at strategic points (nodes).
6 You may want to add another shape such as a triangle to show how many days' duration there are between the main nodes. In this example the piling rigs will arrive and start work on 25 March. They are very expensive to hire therefore cannot be allowed to wait or stand dormant on site.

The piling contractor has to ensure the secant wall is complete and the pile cap cured before the excavation can commence. What then would happen if, say, a hydraulic hose burst on one of the rigs? It may take a morning to replace the hose, test and get the rig back into action therefore the programme may be delayed. The principal contractor would be able to look at the critical path to find out what knock-on effect the delay in piling will have. Would working a Saturday morning enable the schedule to be brought back on time?

Site work is subject to all manner of delays; materials not arriving on time, or being damaged and need replacing etc. Where specialist components are being used, the manufacturer is unlikely to be waiting to make a replacement because an accident on site damaged it beyond repair. The manufacturer will have a 'network programme' for their production. If they are manufacturing, say, washroom pods for another project they may have stock of the components that can be borrowed for the replacement damaged pod. If, say, it was the stone counter top that had been damaged then it may be easier to send the new top and re-assemble on site. As you can see, network diagrams/CPA are essential tools for industry and commerce. They are difficult to produce and need experience to both plan them and use them.

On Ropemaker Asta Powerproject software is used to programme the work. As previously described, planning a Gantt chart is very straightforward and can be easily changed or modified as required. In addition the programme enables each activity to be linked where appropriate to the next critical activity. Other activity bands that are not critical in terms of when they start or complete remain green bands (the colour can be changed though to suit the planners). The critical activities become highlighted with red lines around the green making them stand out on the programme. If a change to the programme is needed the critical path network is electronically linked therefore will change as appropriate. Where the changes prevent the CPA from functioning it becomes apparent on the screen, drawing it to the attention of the planner. The programme enables the planner to link activities from various points along each band providing more accurate detail as to when an activity can or cannot start. All the information is presented on the live screen and with a simple click of the mouse button opens up into yet further detail or can be enlarged for viewing at team meetings.

Meeting rooms on site are equipped with large flat screen TVs and interactive SmartBoards making use of all available technology. This is in contrast with the days when hand coloured bar charts produced in the drawing office as dye line (blue print) and pinned to the site office wall. Although computer software may be considered as expensive, it is essential for 21st century contracts, especially ones of the complexity of the Ropemaker development. Software programmes can be accessed via computer links therefore management can be carried out remotely when required and at any time.

10.13 On-site planning

In parallel with the computerised project planning the planners have to zone areas for working, storing, access routes, and emergency escape. As space is at a premium on site each trade contractor had to arrange delivery of their own materials and plant based on the 'just in time' principle. That means procurement of materials, plant and labour must be programmed with careful planning to ensure materials and plant are not taking up space waiting to be used over long periods. As a guide 2 weeks maximum for storage of materials to maximise the work areas and prevent materials cluttering up the site.

Labour can be programmed more easily as it is mobile. However, materials and plant will need space for offloading and possibly equipment to move it around on site. An 'allocations manager' planner will gather information from the project plan and arrange for the appropriate trade contractor representatives to attend a pre-delivery meeting. Every trade contractor would have been asked how much time and space they would require for example tower crane, lifting jib, gantry and hoist time at an earlier stage. The planner can then allocate lifting time to match the requests, however time allocation is also made on the 'considered time' required by the planner. This is to prevent a trade contractor asking for the tower crane (hook time) all morning to off-load one lorry.

As Ropemaker is a CMA site with CM/TC, it is the responsibility of each trade contractor to order their own materials requirements and schedule deliveries. This is in contrast with standard building contracts where the main contractor orders and finances the project. Waste management is also apart of each trade contractor's remit. The client and principal contractor have an environmental approach to waste disposal on site and provide a range of wheeled 'dump bins' on each floor for specific types of waste from packaging and offcuts etc. Waste is categorised as recyclable (grey, red or green bins) or black for dumping/incineration (see Figure 10.25). The bins are regularly emptied into a similar designated range of skips at ground level.

The project covers the whole site therefore any materials and plant have to be short-term stored on concrete floor decks as they are finished. Standing space for the storage of contractor's materials would then be cordoned off using a series of gates/Herras fencing defining the allocated area. Valuable smaller items would be stored in site safes. The gates/fencing also provide designated pathways for access stopping other contractors moving or trying to store their materials in another contractor's space. The allocated areas are marked on A4 size plans enabling the contractor to identify their areas and the principal contractor to monitor all useable space and to plan for access.

Figure 10.25 Wheeled 'dump bin'.

For example, the curtain walling contractor requires hundreds of completed glazing units to be delivered and stored on site. The programme is to 'weather in' each floor as the building proceeds. ('Weathering in' means to prevent rainwater draining through from the incomplete upper floors and lift shafts. A small water tight step termed a 'bund' is temporarily bonded to the concrete thus preventing any liquids draining off flat areas such as floors and down into the building. All lift shaft and stair openings had the temporary weathering and bunding across them – Figure 10.26.) The delivery vehicles are flat back lorries therefore require mechanical offloading and standing down on pallets. The palletised units must then be taken up to the appropriate floors for storage until required in compounded areas. If, however, the services for example are being run on the same floor, the storage/allocation area will be designed to enable the scissor platforms to operate plus still keep access and emergency routes clear. Mace continually monitor the situation and carry out team meetings monthly, weekly and first thing every morning to ensure that any variations from the plan or allocations can be dealt with. If one trade contractor requires extra hook time or additional storage space there is opportunity to trade off with other trade contractors. The scheme works very well and promotes good working relationships between everyone on site.

So far we have looked at:

1 pre-planning plant and space requirements the trade contractors will require
2 how the Gantt chart compares with the bar chart and why it is more useful
3 how a simple Gantt chart is produced and its limitations
4 the advantages of using a critical path analysis (CPA) and how one would be produced

Bunding bonded to the concrete to form a waterproof seal

Figure 10.26 'Bunding' to prevent ingress of water down the stair wells and lift shafts.

5 the benefits of using computer software such as Asta Powerproject to programme the works on site
6 why it is important to cordon off as opposed to providing compounds on site for space allocations
7 why it is critical to pre-plan and regularly review trade contractor's materials requirements in terms of plant usage, programming for other trades to carry out their work and keep access and emergency evacuation routes clear.

10.14 Topping out

Dating back to medieval times, the clients should provide a celebration when the highest part of the structure is fixed. If the client refused to provide a celebration the tradesman would attach a black flag or material to the highest point to curse the structure. In consequence, some clients have a religious blessing carried out during the topping out ceremony (quite common in far eastern countries).

When the building shell has been completed, a procedure termed 'snagging' will take place. Representatives of the client and principal contractor tour the whole development ensuring the work is to an acceptable standard. At this stage if any defects are found they will be listed as a 'schedule of defects' and presented to the appropriate trade contractors to remedy the situation before the project is repossessed by the client. If the remedial work is not completed before the client takes repossession the trade contractor will require a licence to carry out the work. From the date of handover the client must insure the full project as the principal contractor relinquishes responsibility at that point.

The contract between the landlord British Land Plc. and Mace has been to produce a prestigious office development with some retail areas at ground floor level in the heart of London. The next stage will be the letting agents to introduce clients, most likely blue chip companies, who will lease parts or possibly all of the building from the landlord. They will then enter into new contracts to fit out and furnish offices and retail units in the shell. At the time of completing this book the Ropemaker project was nearing the completion of the external cladding.

Index